Lecture Notes in Mathematics

Edited by A. Dold, Heidelberg and B. Eckmann, Zürich

358

Floyd L. Williams
Massachusetts Institute of Technology, Cambridge, MA/USA

Tensor Products of Principal Series Representations

Reduction of Tensor Products of Principal Series Representations of Complex Semisimple Lie Groups

Springer-Verlag
Berlin · Heidelberg · New York 1973

AMS Subject Classifications (1970): 22-02, 22E45, 43A65, 43A80

ISBN 3-540-06567-9 Springer-Verlag Berlin · Heidelberg · New York
ISBN 0-387-06567-9 Springer-Verlag New York · Heidelberg · Berlin

Offsetdruck: Julius Beltz, Hemsbach/Bergstr.

PREFACE

The aim of these lectures is to study the tensor product of non-degenerate principal series representations of a complex semi-simple Lie group. This will be done from the point of view of the theory of induced representations. We shall observe a delicate relationship between the non-abelian harmonic analysis on the group and the abelian harmonic analysis on its Cartan subgroup--at least in the complex case. The application of the Frobenius-Mackey Reciprocity Theorem, in a sense, brings this relationship to fruition. Ultimately one obtains, thereby, a direct integral reduction of the tensor product. In this reduction only principal series elements occur.

The results presented in these notes encompass earlier results obtained by M. Naimark, [30], G. Mackey, [26], and N. Anh, [1]. The work of these authors focused on the unimodular group only.

We wish to thank Mrs. Lillian White for her skillful typing and preparation of this manuscript. We extend special thanks also to Mrs. Camille Himes who typed revisions of and additions to the original text.

The formulation of Definition 3.5.2 was inspired by conversations with our colleague Professor Edward Wilson. He has often imparted good advice to us for which we are always grateful.

Finally we wish to express indebtedness and very sincere appreciations to Professor Ray A. Kunze. The genesis of these lectures can be traced to the author's thesis which was written under the direction of Professor Kunze and was submitted to the faculty at Washington University in St. Louis, Missouri. At several points Professor Kunze's suggestions proved to be the factor that enabled us to surmount serious obstacles.

F. Williams

Cambridge, Massachusetts
May, 1973

TABLE OF CONTENTS

CHAPTER 1. Introduction 1

CHAPTER 2. Preliminaries on Induced Representations 7

§

2.1 measures on homogeneous spaces 7

2.2 induced representations 13

CHAPTER 3. Representations Induced by Characters of a Subgroup of the Cartan Subgroup 23

§

3.1 some Lie theory 24

3.2 the characters of a Cartan subgroup 33

3.3 the induction of unitary characters of a closed subgroup of a Cartan subgroup 43

3.4 a preliminary theorem 49

3.5 the relative residue of a character and the unitary equivalence of the representations $\operatorname{ind}_{C_1 \uparrow G} \lambda_1$, $\lambda_1 \in \hat{C}_1$.. 59

3.6 $\operatorname{ind}_{C \uparrow G} \lambda$ as a subrepresentation of $L^2(A\backslash G)$ 65

CHAPTER 4. The Tensor Product of Principal Series Representations of a Complex Semi-Simple Lie Group 74

§

4.1 construction of intertwining operators 75

4.2 Harish-Chandra's Plancherel formula 92

4.3 the Frobenius-Mackey reciprocity theorem for non-compact groups 97

4.4 the decomposition of the restriction of the principal series to a Cartan subgroup 107

4.5 the decomposition of the tensor product of principal series representations 115

APPENDIX 121

REFERENCES 130

Chapter 1

Introduction

Let G be a connected complex semi-simple Lie group. Choose an Iwasawa decomposition $G = KAN$ of G where K is a maximal compact subgroup of G, A is abelian, and N is nilpotent. If M is the centralizer of A in K, then $B = MAN$ is a closed minimal parabolic subgroup of G. Let $\hat{B}$ denote the 1-dimensional unitary representations of B. The non-degenerate principal series of unitary representations of G is defined to be the family of induced representations $\operatorname{ind}_{B\uparrow G} \lambda$ where λ varies over $\hat{B}$.

It is now known, by the work of I. Gelfand and M. Naimark, [13], F. Bruhat, [8], K. R. Parthasarathy, R. Ranga-Rao, and V. S. Varadarajan, [31], D. P. Zelobenko, [36], B. Kostant, [18], and N. Wallach, [33], that every non-degenerate principal series representation (for complex G) is irreducible.

We shall be interested in analyzing the tensor product of two principal series representations of an arbitrary complex semi-simple Lie group. This problem has been considered in the special case of the unimodular group $SL(n,\mathbb{C})$ by G. Mackey, [26], and M. Naimark, [30], for $n = 2$, and for arbitrary n by N. Anh, [1]. I. Gelfand and M. Graev, using the method of horospheres, have also studied the tensor product (for the general complex semi-simple Lie group); see [12]. However, many of their arguments seem to be incomplete and somewhat unsatisfactory.

All of our investigations will be carried out within the framework of the theory of induced representations. Let us denote the representation $\operatorname{ind}_{B\uparrow G} \lambda$, $\lambda \in \hat{B}$, by $S(\cdot,\lambda)$. Then, using Bruhat's Lemma and the Mackey theory, one can show that the tensor product $S(\cdot,\lambda_1) \otimes S(\cdot,\lambda_2)$, $\lambda_1, \lambda_2 \in \hat{B}$, itself, is unitarily equivalent to an induced representation. In fact, $S(\cdot,\lambda_1) \otimes S(\cdot,\lambda_2)$ is induced by a certain unitary character of the abelian <u>Cartan subgroup</u> $C = MA$. In Theorem 4.1.10 we construct an explicit intertwining operator $\Phi_{\lambda_1,\lambda_2}$ (and its inverse) which realizes $S(\cdot,\lambda_1) \otimes S(\cdot,\lambda_2)$ as an induced representation. The study of $S(\cdot,\lambda_1) \otimes S(\cdot,\lambda_2)$ is therefore reduced to the study of the family of unitary representations $\operatorname{ind}_{C\uparrow G} \lambda$, where λ varies over the dual group $\hat{C}$. It turns out that the center $\mathbf{Z}(G)$ of G, <u>not the Weyl group</u>, plays a decisive role in the structure of $\operatorname{ind}_{C\uparrow G} \lambda$.

We consider, more generally, the family of (reducible) unitary representations $\operatorname{ind}_{C_1\uparrow G} \lambda_1$, where C_1 is an <u>arbitrary</u> closed subgroup of C. The fundamental property of $\operatorname{ind}_{C_1\uparrow G} \lambda_1$ is that it depends only on the "discrete part" of the unitary character λ_1 of C_1 (see Theorem 3.5.5). Thus, suppose $C_1 = A$ for example (A is the abelian component in the Iwasawa decomposition above).

Then because A is a vector group, the unitary characters of A are indexed by "continuous" parameters only; they have no discrete part. Then Theorem 3.5.5 implies that <u>for any unitary character</u> λ_1 <u>of</u> A

$$\operatorname{ind}_{A\uparrow G} \lambda_1 \simeq \operatorname{ind}_{A\uparrow G} 1$$

<u>where</u> 1 <u>is the trivial character of</u> A (see Corollary 3.5.6).

Another consequence of Theorem 3.5.5 is the following, somewhat surprising, statement about the unitary equivalence of tensor products:

$$S(\cdot,\lambda_1) \otimes S(\cdot,\lambda_2) \simeq S(\cdot,\lambda_3) \otimes S(\cdot,\lambda_4)$$

if and only if

$$\lambda_1 \lambda_2 = \lambda_3 \lambda_4$$

on the center $\mathcal{Z}(G)$ of G , $\lambda_j \in \hat{B}$, $j = 1, 2, 3, 4$; see Theorem 4.1.2. Indeed in the case of principal series representations, if

$$S(\cdot,\lambda) \simeq S(\cdot,\gamma)$$

$\lambda, \gamma \in \hat{B}$, then $\lambda = \gamma$ on $\mathcal{Z}(G)$; but the converse is by no means true.

In a certain sense Theorem 3.5.5 permits a "trivial" analytic continuation of the tensor product $S(\cdot,\lambda) \otimes S(\cdot,\gamma)$ reminiscent of the analytic continuation of the principle series in the classic sense of Kunze and Stein, [19]. In fact, let V be the image of N under the Cartan involution; i.e. V is subgroup of G generated by the negative root vectors. If $H = CV = MAV$ then Kunze, Stein, and E. Wilson, [35], show that for G simply connected,

$$(1.1) \qquad S(\cdot,\lambda)\Big|_H \simeq S(\cdot,\text{res } \lambda)\Big|_H$$

where $\text{res } \lambda$ is the residue of $\lambda \in \hat{B}$; see Definition 3.5.2. This basic fact, together with Bruhat's Lemma, permits a normalization of the principal series. The normalized principal series, in turn, can be analytically continued to yield uniformly bounded representations of G .

Now corresponding to equation (1.1), we show that (without restricting to H)

$$(1.2) \qquad \underset{C\uparrow G}{\text{ind}}\, \lambda \simeq \underset{C\uparrow G}{\text{ind}}(\text{res } \lambda).$$

(1.2) follows from Theorem 3.5.5 by taking $C_1 = C$. Moreover the simply connectivity assumption is not necessary in formulating Definition 3.5.2 or in showing that (1.2) holds.

Theorem 3.5.5 is the most important result in Chapter 3. Another consequence of it is the fact that the tensor product

$S(\cdot,\lambda) \otimes S(\cdot,\gamma)$ can be embedded as a <u>subrepresentation</u> of the <u>right regular representation</u> of G on $L^2(G)$. However, this is not such a useful fact because the embedding is non-canonical. On the other hand, proceeding somewhat more naturally, we characterize $\operatorname{ind}_{C\uparrow G} \lambda$ as a subrepresentation of $L^2(A\backslash G)$. The result, Theorem 3.6.10, depends on the use of the <u>Imprimitivity Theorem</u>.

The problem of decomposing the tensor product as a direct integral is solved in Chapter IV (Theorem 4.5.9). The techniques used by Mackey and Anh for the unimodular group fail in the general case. The reason for this is that their arguments depend on the irreducibility of the restriction of the principal series to a certain parabolic subgroup of $SL(n,\mathbb{C})$. Of course, such good fortune does not prevail in general.

Our idea is to decompose the restriction of $S(\cdot,\lambda)$ to C (Theorem 4.4.13) and then apply the <u>strong form</u> of the Mackey Reciprocity Theorem for non-compact groups; see [26] and see Theorem 4.3.7. We determine the multiplicities encountered in the reduction of $S(\cdot,\lambda_1) \otimes S(\cdot,\lambda_2)$. It results that $S(\cdot,\lambda_1) \otimes S(\cdot,\lambda_2)$ is <u>multiplicity free</u> if and only if N is abelian. This explains, for example, why the principal series of $SL(2,\mathbb{C})$ occur with multiplicity <u>1</u> in the reduction of the tensor product, while on the other hand, the multiplicity "jumps" to $\underline{\infty}$ for the case $SL(3,\mathbb{C})$.

Strictly speaking, the Reciprocity Theorem allows one to write $\operatorname{ind}_{C\uparrow G} \lambda$ as a direct integral for <u>almost all</u> λ in $\hat{C}$.

However, because of Theorem 3.5.5, we can determine the structure of $\operatorname{ind}_{C\uparrow G} \lambda$ for <u>all</u> λ in $\hat{C}$.

The strong form of the Reciprocity Theorem depends ultimately on the so-called <u>weak form</u> of the Reciprocity Theorem; see [26]. Using Harish-Chandra's Plancherel formula for complex semi-simple Lie groups, [23], we shall provide a direct proof of the weak form of the Frobenius-Mackey Reciprocity Theorem; see Theorem 4.3.5.

The assumption that G has a complex structure is basic throughout this memoir.

CHAPTER 2

Preliminaries on Induced Representations

2.1 Measures on homogeneous spaces

The unitary representations which we shall consider are induced in the sense of Mackey, [25]. The construction of such representations involves some basic ideas about measures on homogeneous spaces. Even though these ideas are rather well-known, see [6], [9], [16], or [25], we shall outline them here.

Let G be a locally compact σ-compact group and let H be a closed subgroup of G. Let Δ_G, Δ_H denote the modular functions of G,H, respectively. Let dx, dh denote elements of right Haar measure on G and H respectively. Thus Δ_G satisfies the equations

$$\int_G f(x)dx = \Delta_G(a) \int_G f(a^{-1}x)dx \quad ,$$

$$\int_G f(x)dx = \int_G f(x^{-1})\Delta_G(x)dx$$

for all a in G and for all f in $C_c(G)$, where $C_c(G)$ is the space of complex-valued, continuous, compactly supported functions on G. Corresponding statments about Δ_H are valid as well, of course.

A measure ν on the quotient space $H\backslash G$ (space of right cosets) is <u>G-invariant</u> if

$$\int_{H\backslash G} k(Hxa)d\nu(Hx) = \int_{H\backslash G} k(Hx)d\nu(Hx)$$

for all a in G and for all k in $C_c(H\backslash G)$. Naturally, $H\backslash G$ need not have a G-invariant measure. Such a measure exists if and only if

$$\Delta_G\big|_H = \Delta_H .$$

However, since $H\backslash G$ is paracompact, one can always construct a positive, continuous function ρ on G such that

$$\rho(hx) = \mu(h)\rho(x) \tag{2.1.1}$$

for all (h,x) in $H\times G$ where

$$\mu = \frac{\Delta_G\big|_H}{\Delta_H} .$$

Proposition 2.1.2 <u>Given a positive continuous function</u> ρ <u>on</u> G <u>which satisfies</u> (2.1.1), <u>there exists a unique measure</u> $\nu = \nu(\rho)$ <u>on</u> $H\backslash G$ <u>such that</u>

$$\int_G f(x)\rho(x)dx = \int_{H\backslash G}\int_H f(hx)dh\,d\nu(Hx) \tag{2.1.3}$$

for all f *in* $C_c(G)$. *Moreover*,

$$(2.1.4) \qquad \int_{H\backslash G} k(Hxa)\,d\nu(Hx) = \int_{H\backslash G} k(Hx)\,\frac{\rho(xa^{-1})}{\rho(x)}\,d\nu(Hx)$$

for all a *in* G *and* k *in* $C_c(H\backslash G)$.

We shall give a proof of Proposition 2.1.2 which is based on the following

Lemma 2.1.5 *If* f *is a function in* $C_c(G)$ *such that*

$$\int_H f(hx)\,dh = 0$$

for all x *in* G, *then*

$$\int_G f(x)\rho(x)\,dx = 0.$$

Proof: Choose f_1 in $C_c(G)$ such that

$$\int_H f_1(hx)\,dh = 1$$

for all x in the support of f. Define $f_2 = f_1\rho$; f_2 belongs to $C_c(G)$. We have

$$0 = \int_G \int_H f_2(x) f(hx)\,dh\,dx$$

$$= \int_H \int_G f_2(x) f(hx)\,dx\,dh$$

$$= \int_H \int_G f_2(h^{-1}x)f(x)\Delta_G(h)dxdh$$

$$= \int_G \int_H f_2(h^{-1}x)f(x)\Delta_H(h) \frac{\rho(hx)}{\rho(x)} dhdx,$$

by (2.1.1)

$$= \int_G \int_H f_2(hx)f(x)\Delta_H(h^{-1}) \frac{\rho(h^{-1}x)}{\rho(x)} \Delta_H(h)dhdx$$

$$= \int_G \int_H f_1(hx)\rho(hx)f(x)\mu(h^{-1})dhdx$$

$$= \int_G \int_H f_1(hx)\rho(x)f(x)dhdx,$$

by (2.1.1)

$$= \int_G f(x)\rho(x)dx,$$

since

$$f(x)\rho(x) \int_H f_1(hx)dh = f(x)\rho(x)$$

for all x in G. The lemma is therefore proved. The proof of Proposition 2.1.2 goes as follows:

Take k in $C_c(H\backslash G)$. Choose f in $C_c(G)$ such that

$$k(Hx) = \int_H f(hx)dh \tag{2.1.6}$$

for all x in G. Set

$$L(k) = \int_G f(x)\rho(x)dx \quad . \tag{2.1.7}$$

By Lemma 2.1.5 equation (2.1.7) defines a positive linear functional L on $C_c(H\backslash G)$; i.e. L is independent of the particular function f which represents k in (2.1.6). By the Riesz representation theorem, [16], there is a unique Radon measure $\nu = \nu(\rho)$ on $H\backslash G$ such that

$$L(k) = \int_{H\backslash G} k(Hx)d\nu(Hx)$$

for all k in $C_c(H\backslash G)$; i.e. ν satisfies (2.1.3).

Finally

$$\int_{H\backslash G} k(Hxa)d\nu(Hx)$$

$$= \int_{H\backslash G}\int_H f(hxa)dhd\nu(Hx)$$

$$= \int_G f(xa)\rho(x)dx,$$

by (2.1.3)

$$= \int_G f(x)\frac{\rho(xa^{-1})}{\rho(x)}\rho(x)dx$$

$$= \int_{H\backslash G}\int_G f(hx)\frac{\rho(hxa^{-1})}{\rho(hx)}dhd\nu(Hx),$$

by (2.1.3)

$$= \int_{H\backslash G} \kappa(Hx) \frac{\rho(xa^{-1})}{\rho(x)} \, d\nu(Hx),$$

by (2.1.1) and (2.1.6). This proves (2.1.4).

Note that (2.1.1) implies

$$(2.1.8) \qquad \rho_a : x \to \frac{\rho(xa^{-1})}{\rho(x)}, \quad x \text{ in } G,$$

is indeed a function on $H\backslash G$, for all a in G. Equation (2.1.4) means that ν, though not G-invariant, is <u>quasi-invariant</u>; i.e. for all a in G the right translate ν_a of ν by a has the same null sets as ν. The function ρ_a in (2.1.8) is the Radon-Nikodym derivative of ν_a with respect ν.

We observe moreover that if $\Delta_G|_H = \Delta_H$, in particular, then the function $\rho \equiv 1$ satisfies (2.1.1) and (2.1.4) means precisely that the corresponding measure ν, given by Proposition 2.1.2, is G- invariant.

In other important special cases, solutions of (2.1.1) are obtained directly, as well. For example, suppose G is a connected semi-simple Lie group. Choose an Iwasawa decomposition

$$G = ANK$$

of G where A, N, K are closed subgroups of G, A is abelian and N is nilpotent. Take $H = MAN$ where M is the centralizer of A in K. G is known to be unimodular, i.e. $\Delta_G \equiv 1$, and H is known to be a closed subgroup of G. We can construct ρ (which satisfies (2.1.1)) for the pair G,H by setting

$$\rho(x) = \frac{1}{\Delta_H\big(a(x)\big)}$$

for x in G, where $a(x) \in A$ is the A component of x.

2.2 Induced representations

Suppose G, H, ρ are as above, where ρ satisfies (2.1.2). Let λ be a unitary representation of H on a Hilbert space K_λ; all representations are assumed to be continuous. Following Mackey, [25], one can construct a unitary representation of G as follows: Define $H(\lambda)$ to be the space of vector-valued measurable functions

$$f\colon\ G \to K_\lambda$$

from G to K_λ such that

(i) $f(hx) = \lambda(h)f(x)$ for all (h,x) in $H \times G$

(ii) $\displaystyle\int_{H\backslash G} ||f(x)||^2 \, d\nu(Hx) < \infty$.

Since λ is unitary (i) implies that

$$x \to ||f(x)||^2 \quad , \quad x \in G \quad ,$$

is a function on $H\backslash G$. Therefore, condition (ii) is well-defined. $H(\lambda)$ modulo the null space of functions is a Hilbert space which we also denote by $H(\lambda)$. Now (2.1.4) implies immediately that the equation

$$(2.2.1) \qquad \left(T(a,\lambda)f\right)(x) = \left[\frac{\rho(xa)}{\rho(x)}\right]^{\frac{1}{2}} f(xa) \quad ,$$

$(a,x,f) \in G\times G\times H(\lambda)$, defines a unitary representation

$$T(\cdot,\lambda): \quad a \to T(a,\lambda)$$

of G on $H(\lambda)$. $T(\cdot,\lambda)$ is called the representation of G induced by the unitary representation λ of H . We shall also denote $T(\cdot,\lambda)$ by $\operatorname{ind}_{H \uparrow G} \lambda$.

In [3], Blattner gives an alternative construction of induced representations which does not rely on the existence of quasi-invariant measures or even the assumption that G is separable.

Definition 2.2.2 Let $\mathfrak{M}$ be a σ-algebra of subsets of a set X. A spectral measure on X is a mapping P with domain $\mathfrak{M}$, whose values $P(\sigma)$, $\sigma \in \mathfrak{M}$, are projections on a (fixed) Hilbert space $\mathcal{H}$ such that

(i) $P(\emptyset) = 0$, $P(X) = I$

(ii) $P(\sigma_1 \cap \sigma_2) = P(\sigma_1)P(\sigma_2) = P(\sigma_2)P(\sigma_1)$ for σ_1, σ_2 in $\mathfrak{M}$

(iii) $P\left(\bigcup_{i=1}^{\infty} \sigma_i\right)x = \sum_{i=1}^{\infty} P(\sigma_i)x$ for all x in $\mathcal{H}$ whenever $\{\sigma_i\}_{i=1}^{\infty}$ is a pairwise disjoint sequence in $\mathfrak{M}$.

Actually, by some perseverance, one can deduce (ii) from (i) and (iii). We shall simply assume (ii) as part of the definition.

There is a canonical spectral measure $P(\cdot,\lambda)$ associated with any induced representation $T(\cdot,\lambda)$. $P(\cdot,\lambda)$ is defined as follows: Let $\mathbb{B}(H\backslash G)$ denote the σ-algebra of Borel subsets of $H\backslash G$. To each Borel set σ in $\mathbb{B}(H\backslash G)$ we assign a projection $P(\sigma,\lambda)$ on $H(\lambda)$, the induced representation space, by setting

$$P(\sigma,\lambda)f = (\chi_\sigma \circ \Pi)f \quad , \tag{2.2.3}$$

where f belongs to $H(\lambda)$,

$$\Pi: \quad G \to H\backslash G$$

is the natural projection, and x_σ is the characteristic function of σ.

Proposition 2.2.4 *The induced representation* $T(\cdot,\lambda)$ *and the canonical spectral measure* $P(\cdot,\lambda)$ *satisfy the following important formula:*

$$(2.2.5) \qquad T(a,\lambda)P(\sigma,\lambda) = P(\sigma a^{-1},\lambda)T(a,\lambda)$$

for all a *in* G *and for all Borel subsets* σ *of* $H\backslash G$.

Formula (2.2.5) is an immediate consequence of (2.2.1) and (2.2.3). *The existence of a spectral measure* P *which commutes with a unitary representation* T *of* G *in the sense of* (2.2.5) *actually characterizes* T *as an induced representation.* This is the essential content of the *Imprimitivity Theorem* which we shall formulate more precisely, presently.

If $T(\cdot,\lambda)$ is irreducible, then necessarily λ is irreducible. This is easy to see because if a bounded operator A commutes with each $\lambda(h)$, h in H, then A induces a bounded operator $\tilde{A}$ on $H(\lambda)$ which commutes with each $T(a,\lambda)$, a in G. The irreducibility of λ, for irreducible $T(\cdot,\lambda)$, also follows from the fact that induction and direct summation commute. More precisely if $\{\lambda_\alpha\}$ is a family of unitary representations of H then

$$\underset{H\uparrow G}{\text{ind}} \sum_{\alpha} \oplus \lambda_\alpha \simeq \sum_{\alpha} \oplus \underset{H\uparrow G}{\text{ind}} \lambda_\alpha \tag{2.2.6}$$

where $\simeq$ denotes unitary equivalence. We shall omit the (easy) proof of (2.2.6) since we shall not make use of it. We shall need the following deeper result, however, which is due to Mackey:

Theorem 2.2.7 *Let* G *be a locally compact group and let* H *be a closed subgroup of* G. *Given a unitary representation* λ *of* H, *let* $T(\cdot,\lambda)$ *be the induced representation of* G *and let* $P(\cdot,\lambda)$ *be the canonical spectral measure on* $H\backslash G$. *Then* λ *is irreducible if and only if the pair* $\bigl(T(\cdot,\lambda),\ P(\cdot,\lambda)\bigr)$ *is irreducible. Moreover if* γ *is another unitary representation of* H *then*

$$\gamma \simeq \lambda$$

if and only if

$$\bigl(T(\cdot,\lambda), P(\cdot,\lambda)\bigr) \simeq \bigl(T(\cdot,\gamma), P(\cdot,\gamma)\bigr) .$$

Blattner has presented a beautiful, somewhat algebraic, proof of Theorem 2.2.7 in [4]. Unfortunately, the ideas are too involved to be included here.

Of central importance in the theory of induced representations is the problem of analyzing the structure of the restriction of an induced representation to a closed subgroup. The basic general theorem for doing this is the Mackey Subgroup Theorem, [25]. We shall give an explicit solution of this problem in the case of the restriction of a non-degenerate principal series representation to a Cartan subgroup; see Theorem 4.4.13. Therefore we will not make use of the Subgroup Theorem.

Theorem 2.2.8 *Let* H_1 , H_2 *be closed subgroups of a locally compact group* G. *Assume that there are at most countable many* $(H_1 : H_2)$ *double cosets. Let* λ_1 , λ_2 *be unitary representations of* H_1 , H_2 , *respectively. For each element* (x,y) *in* $G \times G$, *let* λ_1^x , λ_2^y *be the unitary representations of* $x^{-1}H_1x$, $y^{-1}H_2y$, *respectively, defined by*

$$\lambda_1^x(h_1) = \lambda_1(xh_1x^{-1})$$

$$\lambda_2^y(h_2) = \lambda_2(yh_2y^{-1})$$

where $(h_1,h_2) \in (x^{-1}H_1x) \times (y^{-1}H_2y)$. *Let*

$$T^{x,y} = \lambda_1{}^x\Big|_{H_{x,y}} \otimes \lambda_2{}^y\Big|_{H_{x,y}} ,$$

where

$$H_{x,y} = x^{-1}H_1x \cap y^{-1}H_2y .$$

Let

$$V^{x,y} = \operatorname*{ind}_{H_{x,y}\uparrow G} T^{x,y} .$$

Then, to within unitary equivalence, $V^{x,y}$ *only depends on the double coset* $H_1xy^{-1}H_2$ *containing* xy^{-1} *so that we may write*

$$V^{x,y} = V^d$$

where $d = H_1xy^{-1}H_2$. *Moreover*

$$\operatorname*{ind}_{H_1\uparrow G} \lambda_1 \otimes \operatorname*{ind}_{H_2\uparrow G} \lambda_2 \simeq \sum_d \otimes V^d \tag{2.2.9}$$

where the sum is taken over those double cosets d *which have positive* G-*Haar measure.*

We shall *not* need Theorem 2.2.8. We shall *illustrate* an application of it in Chapter 4 in the case $H_1 = H_2$ is a minimal parabolic subgroup. In fact, in this case, we shall show how to construct the unitary equivalence in (2.2.9) quite directly; see Theorem 4.1.10.

Earlier we observed that to every induced representation $T(\cdot,\lambda)$ there is an associated spectral measure $P(\cdot,\lambda)$ such that (2.2.5) holds. Equation (2.2.5) is the motivation for

Definition 2.2.10 Let G be a locally compact group which acts (say on the right) continuously on a locally compact space S, and let T be a unitary representation of G on a Hilbert space H_T. A spectral measure P on S, defined on the Borel subsets of S, which takes values in the set of projections on H_T is called a system of imprimitivity for T, based on the action of G on S, if

$$T(a)P(\sigma) = P(\sigma a^{-1})T(a)$$

for all a in G and for all Borel subsets σ of S.

Thus, by (2.2.5), if H is a closed subgroup of G and if λ is a unitary representation of H, then $P(\cdot,\lambda)$ is a (canonical) system of imprimitivity for $T(\cdot,\lambda)$ based on the (transitive) action of G on $H\backslash G$. Conversely, we have the celebrated Imprimitivity Theorem:

Theorem 2.2.11 *Let* G *be a locally compact group and let* T *be a unitary representation of* G. *Suppose there is a system of imprimitivity* P *for* T *based on the action of* G *on* $H\backslash G$, *where* H *is a closed subgroup of* G. *Then there exists a unitary representation* λ *of* H, *unique to within unitary equivalence, such that*

$$(T,P) \simeq \bigl(T(\cdot,\lambda),P(\cdot,\lambda)\bigr)$$

Thus, in particular, T is unitarily equivalent to an induced representation. λ is necessarily unique (to within equivalence) because of Theorem 2.2.7.

The Imprimitivity Theorem is due to Mackey, [28], although it was first proved by Frobenius in the very special case of a finite group G. Frobenius also gave a somewhat different formulation of it. It is not possible to give a proof of Theorem 2.2.11 here since this would lead us well beyond the scope of this work. Mackey's original proof is very difficult and measure-theoretic. A nicer proof by Blattner, [5], is available; also see Loomis, [20].

The idea of the proof is to convert $C_c\bigl((H\backslash G)\times G\bigr)$ into a convolution algebra and then show that representations of this algebra correspond to pairs (T,P) where P is a system of imprimitivity for T. One defines convolution and an adjoint operation $*$ by setting

$$(f * g)(Ha,x) = \int_G f(Ha,y)\, g(Hay,y^{-1}x)\, dy$$

and

$$f^*(Ha,x) = \overline{f}(Hax,x^{-1})\, \Delta_G(x^{-1}) \quad ,$$

where f , $g \in C_c\big((H\backslash G)\times G\big)$, $(a,x) \in G \times G$ and dy now denotes left Haar measure on G.

CHAPTER 3

Representations Induced by Characters of a Subgroup of the Cartan Subgroup

In this chapter we consider the family of representations of a complex semi-simple Lie group G induced by unitary characters of a closed subgroup C_1 of a Cartan subgroup C of G .

Suppose we denote this family by $\mathcal{J}_{C_1}$. Then $\mathcal{J}_{C_1}$ includes the family of tensor products of non-degenerate principal series representations of G in the special case that $C_1 = C$; cf. Chapter 1. Indeed for this reason alone we are motivated to study $\mathcal{J}_{C_1}$. The basic facts concerning the family $\mathcal{J}_{C_1}$ are given in Theorem 3.5.5.

We begin by sketching the Lie Theory that is pertinent for later discussions.

3.1 Some Lie theory

Let G be a connected semi-simple Lie group and let $\mathfrak{g}$ be the Lie algebra of G over the real field $\mathbb{R}$. We shall assume that G is complex; i.e. we assume that $\mathfrak{g}$ has a complex structure J. This means that J is an $\mathbb{R}$-endomorphism of the $\mathbb{R}$-vector space $\mathfrak{g}$ and J satisfies

(3.1.1) $$J^2 = -I$$

(3.1.2) $$J \circ \mathrm{ad}_X = \mathrm{ad}_X \circ J$$

for all X in $\mathfrak{g}$, where ad is the adjoint representation of $\mathfrak{g}$ on $\mathfrak{g}$.

By (3.1.1) we can define complex scalar multiplication on $\mathfrak{g}$ by setting

$$cX = aX + bJX ,$$

where $c = a + \sqrt{-1}\, b \in \mathbb{C}$. With multiplication by complex scalars and with the bracket operation on $\mathfrak{g}$, $\mathfrak{g}$ becomes a $\mathbb{C}$-Lie algebra $\mathfrak{g}^J$, since (3.1.2) and (3.1.1) guarantee that the bracket will be $\mathbb{C}$-bilinear. The Killing forms $B_{\mathfrak{g}}$, $B_{\mathfrak{g}^J}$ on $\mathfrak{g}$, $\mathfrak{g}^J$ are related by the formula

$$B_{\mathfrak{g}}(X,Y) = 2 \operatorname{Re} B_{\mathfrak{g}^J}(X,Y)$$

for all X, Y in $\mathfrak{g}$. In particular $\mathfrak{g}^J$ is also semi-simple.

Choose a compact real form $\mathfrak{k}$ of $\mathfrak{g}^J$. Then

(3.1.3) $$\mathfrak{g} = \mathfrak{k} \oplus J\mathfrak{k}$$

is a Cartan decomposition of $\mathfrak{g}$; i.e.

$$B\big|_{\mathfrak{k}} \text{ is strictly negative definite,}$$

$$B\big|_{J\mathfrak{k}} \text{ is strictly positive definite,}$$

and the map

$$\theta : X + JY \longrightarrow X - JY, \tag{3.1.4}$$

$X, Y \in \mathfrak{k}$, is an $\mathbb{R}$-automorphism of $\mathfrak{g}$. θ is called the <u>Cartan involution</u>; see [15].

Choose a maximal abelian subspace $\mathfrak{a}$ of $J\mathfrak{k}$ and put

$$\mathfrak{h} = \mathfrak{a} + J\mathfrak{a} \qquad \left(\mathfrak{a} \cap J\mathfrak{a} = \{0\}\right).$$

Then $\mathfrak{h}$ is J-invariant; i.e. J induces a complex structure J on $\mathfrak{h}$. Moreover $\mathfrak{h}^J$ is a Cartan subalgebra of $\mathfrak{g}^J$. $\mathfrak{g}^J$, $\mathfrak{h}^J$ will also be denoted by $\mathfrak{g}$, $\mathfrak{h}$, respectively. The set of non-zero roots of $\mathfrak{g}$ relative to $\mathfrak{h}$ will be denoted by Δ. Δ^+ will denote the set of positive roots relative to some lexicographic ordering on $\mathfrak{a}$ (or the real dual of $\mathfrak{a}$), and

$$\Pi = \{\alpha_1, \cdots, \alpha_\ell\}$$

will denote a system of <u>simple</u> roots, where

$$\ell = \dim_{\mathbb{C}} \mathfrak{h}.$$

For each $\alpha \in \Delta$, choose $H_\alpha \in \mathfrak{h}$ so that

$$B(H, H_\alpha) = \alpha(H)$$

for all $H \in \mathfrak{h}$, where $B = B_{\mathfrak{g}^J}$ is the Killing form on $\mathfrak{g}^J$; recall that $B|_{\mathfrak{h}^J}$ is non-degenerate. Define

(3.1.5)
$$\mathfrak{h}_{\mathbb{R}} = \mathbb{R} H_\alpha, \ \alpha \in \Delta$$
$$\mathfrak{h}^*_{\mathbb{R}} = \mathbb{R} \Delta \ .$$

Then, it is known,

$$\mathfrak{a} \equiv \mathfrak{h}_{\mathbb{R}},$$

and the roots are <u>real valued</u> on $\mathfrak{h}_{\mathbb{R}}$.

Define

(3.1.6)
$$\mathfrak{n} = \sum_{\alpha \in \Delta^+} \mathfrak{g}_\alpha,$$

(3.1.7)
$$\mathfrak{v} = \sum_{\alpha \in \Delta^+} \mathfrak{g}_{-\alpha},$$

where $\mathfrak{g}_\alpha$ is the (1-dimensional) root space corresponding to $\alpha \in \Delta$. $\mathfrak{n}$ and $\mathfrak{v}$ are nilpotent subalgebras of $\mathfrak{g}^J$, and $\mathfrak{v} = \theta\mathfrak{n}$, $\mathfrak{n} = \theta\mathfrak{v}$. Let K, A, N, V, M, C denote the Lie subgroups of G whose corresponding Lie algebras are $\mathfrak{k}$, $\mathfrak{a}$, $\mathfrak{n}$, $\mathfrak{v}$, $J\mathfrak{a} = \sqrt{-1}\,\mathfrak{a}$, $\mathfrak{h}$, respectively. Then M is compact, K is a maximal compact subgroup, A, N, V, C are closed, A, M, C are abelian, N, V are nilpotent, A, N, V are simply connected, M is the centralizer of A in K, and C = MA is a Cartan subgroup of G. Moreover

(3.1.8)
$$G = KAN$$

is an Iwasawa decomposition of G, where AN is a closed solvable subgroup. Also

(3.1.9)
$$B = MAN$$

is a closed solvable subgroup of G such that BV is a dense, open submanifold of G whose complement has Haar measure zero; see [19] for a proof.

Let M' be the normalizer of A in K. Then

$$W = M'/M \tag{3.1.10}$$

is the (finite) Weyl group. The fact that $G - BV$ has measure zero is a consequence of Bruhat's Lemma, [24], which we state as follows:

THEOREM 3.1.11 <u>Every</u> $B:B$ <u>double coset</u> (or orbit of B <u>in</u> $B\backslash G$) <u>contains exactly one coset of</u> W. <u>Thus the</u> $B:B$ <u>double cosets are finite in number and are in one-one correspondence with</u> W. <u>In addition, there exists a unique double coset</u> Bp_0B, $p_0 \in M'$, <u>such that</u> Bp_0B <u>has positive</u> G-<u>Haar measure</u>.

Bruhat's Lemma is valid for real semi-simple groups, independently of our assumption that G is complex.

Let Ad be the adjoint representation of G on $\mathfrak{g}$: Ad(x) is the automorphism of $\mathfrak{g}$ corresponding to the inner automorphism $y \longrightarrow xyx^{-1}$ of G. If $p \in M'$, then it follows easily that

$$\mathrm{Ad}(p)\,\mathfrak{a} = \mathfrak{a}$$

and since J commutes with Ad, we get

$$\mathrm{Ad}(p)\,\mathfrak{h} = \mathfrak{h}. \tag{3.1.12}$$

Since M is the centralizer of A in K, Ad(p) H only depends on the coset in W which contains p, for $H \in \mathfrak{h}$. In other words (3.1.12) means that W acts (linearly) on $\mathfrak{h}$ and, hence, passing to the contragradient representation, W acts on the complex dual $\mathfrak{h}^*$. We shall write

$$\mathrm{Ad}(p)\xi = p\xi$$

for $p \in W$ (or M') and $\xi \in \mathfrak{h}^*$. Thus

(3.1.13) $$(p\xi)(H) = \xi\left(\mathrm{Ad}(p^{-1})H\right)$$

for all H in $\mathfrak{h}$ and $\xi \in \mathfrak{h}^*$.

Given α in Δ, an element p_α in M' can be chosen, once and for all, so that

(3.1.14) $$p_\alpha \xi = \xi - \frac{2(\xi|\alpha)\alpha}{(\alpha|\alpha)}$$

for all ξ in $\mathfrak{h}^*$, where $(\cdot|\cdot)$ is the symmetric bilinear form on $\mathfrak{h}^*$ induced by the Killing form $B_{\mathfrak{g}^J}$. The <u>simple Weyl reflections</u> p_α, $\alpha \in \Pi$, generate W and, concerning the element p_0 in Theorem 3.1.11, there exists a permutation $\left(n(1), \cdots, n(\ell)\right)$ of the letters $1, 2, \cdots, \ell$, $\ell = \dim_{\mathbb{C}} \mathfrak{h}$, such that

(3.1.15) $$p_0\alpha_j = -\alpha_{n(j)}, \quad 1 \le j \le \ell.$$

We also have

(3.1.16) $$V = p_0 N p_0^{-1}.$$

Lemma 3.1.17 Given $\alpha_j \in \Pi$, let p_{α_j} be the corresponding simple Weyl reflection which satisfies (3.1.14). If $\alpha \in \Delta^+$ such that $\alpha \neq \alpha_j$, then $p_{\alpha_j}\alpha \in \Delta^+$. Of course $p_{\alpha_j}\alpha_j = -\alpha_j$.

Proof: Using (3.1.13) it is easy to see that $p\Delta = \Delta$ for any $p \in M'$ so $p_{\alpha_j}\alpha$ is a root. Since $\alpha > 0$, and since $\alpha \neq \alpha_j$, write $\alpha = \sum_{i=1}^{\ell} k_j \alpha_j$ where each k_j is a non-negative integer with $k_{i_o} > 0$ for at least one $i = i_o$, $i \neq j$. By (3.1.14)

$$p_{\alpha_j}\alpha = \sum_{i \neq j} k_i \alpha_i + (k_j - n_j)\alpha_j ,$$

where $n_j = \dfrac{2(\alpha|\alpha_j)}{(\alpha_j|\alpha_j)}$. Actually n_j is an integer, although we do not need this fact. If $p_{\alpha_j}\alpha < 0$ we would have

$$p_{\alpha_j}\alpha = \sum_{i=1}^{\ell} m_i \alpha_i ,$$

where each integer $m_i \leq 0$. Since Π is an $\mathbb{R}$-basis of $\mathfrak{h}^*_{\mathbb{R}}$ (see (3.1.5)), this would force $m_j = k_i$ for all $i \neq j$ so that $m_{i_o} = k_{i_o} > 0$ would give a contradiction. Since $\alpha \neq 0$, $p_{\alpha_j}\alpha \neq 0$. Therefore $p_{\alpha_j}\alpha > 0$. This concludes the proof.

Given $p \in M'$, define

$$\mathfrak{n}_p = \sum_{\substack{\alpha > 0 \\ p_\alpha < 0}} \mathfrak{g}_\alpha$$

(3.1.18)

$$\mathfrak{n}'_p = \sum_{\substack{\alpha > 0 \\ p_\alpha > 0}} \mathfrak{g}_\alpha$$

Then $\mathfrak{n} = \mathfrak{n}_p + \mathfrak{n}'_p$, $\mathfrak{n}_p \cap \mathfrak{n}'_p = \{0\}$. $\mathfrak{n}_p, \mathfrak{n}'_p$ are subalgebras of $\mathfrak{n}$. Define N_p , N'_p to be the Lie subgroups of N corresponding to $\mathfrak{n}_p, \mathfrak{n}'_p$, respectively.

THEOREM 3.1.19 N_p, N'_p are closed, simply connected subgroups of N such that

$$N = N_p N'_p$$

and $N_p \cap N'_p = \{e\}$, e being the identity element of G .

Analogous statements hold for V . A proof of Theorem 3.1.19 is given in [19].

Later on we shall use the Fourier transform on $\mathbb{C}$ to analyze certain induced representations. Theorem 3.1.19 will be used in the special case where p is a simple Weyl reflection, say p_{α_j} , $\alpha_j \in \Pi$. In this case, we have

$$(3.1.20) \qquad \mathfrak{n}_{p_{\alpha_j}} \equiv \mathfrak{g}_{\alpha_j} ,$$

by Lemma 3.1.17. Therefore $\mathfrak{n}_{p_{\alpha_j}}$ and $N_{p_{\alpha_j}}$ are 1-dimensional and if $X_j \in \mathfrak{g}_{\alpha_j} - \{0\}$, the map

$$(3.1.21) \qquad z \longrightarrow \exp z X_j$$

is an analytic isomorphism of $\mathbb{C}$ onto $N_{p_{\alpha_j}}$.

The results we obtain in later sections, then, will rely heavily upon the assumption that G has a complex structure. In the general real case, the root spaces (i.e., the so-called restricted root spaces) attain dimensions greater than one.

Elements in the set BV are decomposed uniquely (and continuously) into B and V components: If $a \in BV$ we shall write

$$(3.1.22) \qquad a = b(a)\, v(a) .$$

Moreover, $B = CN$ so we shall also write

$$(3.1.23) \qquad b = c(b)\, n(b)$$

for $b \in B$; $B \cap V = C \cap N = \{e\}$, where e is the identity in G ;

$$\big(b(a)\,,\ v(a)\,,\ c(b)\,,\ n(b)\big) \in B \times V \times C \times N .$$

Consistent with the "decomposition"

$$G = BV \qquad \text{(almost everywhere)},$$

the right Haar measures dx, db, dv, dc, dn on G, B, V, C, N respectively are related by the following well-known formulas :

$$\text{(3.1.24)} \qquad \int_G f(x)\,dx = \int_V \int_B f(bv)\,\mu(b^{-1})\,db\,dv$$

$$= \int_N \int_V \int_C f(cnv)\,dc\,dv\,dn$$

for any continuous f on G with compact support in BV, where

$$\mu = \frac{\Delta_G|_B}{\Delta_B} .$$

(3.1.24) is proved in [19], for example. We shall need the formulas:

$$\text{(3.1.25)} \qquad \mu(c) = \exp\left[2\sigma\,(\mathrm{Re}\,\log c)\right]$$

for $c = \exp H \in C$, where $\log c = H = H_1 + \sqrt{-1}\,H_2 \in \mathfrak{a} + \sqrt{-1}\,\mathfrak{a} = \mathfrak{h}$, $\mathrm{Re}\,H = H_1$, and $\sigma = \sum_{\alpha \in \Delta^+} \alpha$

$$\text{(3.1.26)} \qquad \mu\left(p_o^{-1}\,cp_o\right) = \mu(c^{-1})$$

for all $c \in C$.

3.2 The characters of a Cartan subgroup

Now it is convenient to consider the characters (not necessarily unitary) of C . Their parametrization is well known and is outlined here, briefly.

Since $C = AM$, $M \cap A = \{e\}$, e being the identity of G , the characters of C are determined by those of M and A . The characters of M are necessarily unitary, M being compact, and they are identified with the set of weights

$$(3.2.1) \quad \hat{M} = \{\eta \in \mathfrak{h}^*_{\mathbb{R}} \mid \eta(H) \in 2\pi\sqrt{-1}\,\mathbb{Z} \text{ for every } H \in L_e\}$$

of $\sqrt{-1}\,\mathfrak{a}$, where $\mathbb{Z}$ is the ring of integers and

$$(3.2.2) \quad L_e = \{H \in \sqrt{-1}\,\mathfrak{a} \mid \exp H = e\}$$

is the unit lattice in $\mathfrak{h}$. The exponential mapping of $\mathfrak{g}$ into G takes $\sqrt{-1}\,\mathfrak{a}$ onto M , $\mathfrak{a}$ onto A , and onto C , and we have

$$(3.2.3) \quad \eta(\exp \sqrt{-1}\,H) = \exp \sqrt{-1}\,\eta(H)$$

for any character η of M , where $(\eta, H) \in \hat{M} \times \mathfrak{a}$.

The characters ξ of the (simply connected) group A are parametrized by $\mathfrak{h}^*$ and we have

$$(3.2.4) \quad \xi(\exp H) = \exp \xi(H) ,$$

where

$$(\xi, H) \in \mathfrak{h}^* \times \mathfrak{a} .$$

From (3.2.3) and (3.2.4) we can identify the characters λ

of C with $\mathfrak{h}^* \times \hat{M}$, where we write

(3.2.5) $$\lambda = (\xi, \eta)$$

to mean that

(3.2.6) $$\lambda(c) = e^{\xi(H_1) + \sqrt{-1}\,\eta(H_2)}$$

whenever

$$c = \exp H \in C \quad , \quad H = H_1 + \sqrt{-1}\,H_2 \in \mathfrak{a} + \sqrt{-1}\,\mathfrak{a} = \mathfrak{h} \quad .$$

Under this identification, the unitary characters, $\hat{C}$, correspond to $\sqrt{-1}\,\mathfrak{h}^*_{\mathbb{R}} \times \hat{M}$.

Let us denote the center of G by $\mathbf{Z}(G)$. $\mathbf{Z}(G)$ is a finite group, as G is complex, and

$$\mathbf{Z}(G) \subset M \quad .$$

We want to give a necessary and sufficient condition that two characters of C will coincide on $\mathbf{Z}(G)$.

Since G is connected, $\mathbf{Z}(G)$ is the kernel of the adjoint representation Ad of G on $\mathfrak{g}$. We define

(3.2.7) $$L_{\mathbb{Z}} = \mathbb{Z}\Pi = \left\{ \Sigma\, m_j \alpha_j \,\middle|\, m_j \in \mathbb{Z} \ , \ \alpha_j \in \Pi \right\} \quad ;$$

$L_{\mathbb{Z}}$ is the <u>integer lattice</u> in $\mathfrak{h}^*$.

Proposition 3.2.8 <u>If</u> $\alpha \in \Delta$ <u>and</u> $X_\alpha \in \mathfrak{g}_\alpha$, <u>then</u>

$$\mathrm{Ad}(c)\, X_\alpha = e^{\alpha(H)}\, X_\alpha$$

for $c = \exp H \in C$, $H \in \mathfrak{h}$.

Proof: We have

$$\mathrm{Ad}(c) X_\alpha = \mathrm{Ad}(\exp H) X_\alpha = e^{\mathrm{ad}\, H} X_\alpha = \left(I + \mathrm{ad}\, H + \frac{\mathrm{ad}\, H^2}{2!} + \cdots \right) X_\alpha$$

$$= X_\alpha + \alpha(H) X_\alpha + \frac{\alpha(H)^2}{2!} X_\alpha + \cdots = e^{\alpha(H)} X_\alpha \ .$$

A useful consequence of Proposition 3.2.8 is

Proposition 3.2.9 If $\exp \sqrt{-1}\, H \in \mathbb{Z}(G)$, where $H \in \mathfrak{a}$, then for every $\alpha \in \Delta$, $\alpha(H) \in 2\pi \mathbb{Z}$.

Proof: Let $c = \exp\sqrt{-1}\, H$ and take $X_\alpha \in \mathfrak{g}_\alpha - \{0\}$. Since $c \in \mathbb{Z}(G)$ and $\mathbb{Z}(G)$ is the kernel of Ad , we get

$$X_\alpha = \mathrm{Ad}(c) X_\alpha = e^{\sqrt{-1}\, \alpha(H)} X_\alpha \ ,$$

by Proposition 3.2.8, so $X_\alpha \neq 0$ implies

$$e^{\sqrt{-1}\, \alpha(H)} = 1$$

and since the roots are real-valued on $\mathfrak{a}$, it follows that $\alpha(H) \in 2\pi Z$.

Corollary 3.2.10 $L_{\mathbb{Z}} \subset \hat{M}$; i.e. $L_{\mathbb{Z}}$ is an abelian subgroup of $\hat{M}$.

Proof: Each α_j (= simple root) $\in \mathfrak{h}^*_{\mathbb{R}}$ and if $\sqrt{-1}\, H \in L_e$, $H \in \mathfrak{a}$, then

$$\exp\sqrt{-1}\, H = e \in \mathbb{Z}(G)$$

so $\alpha(\sqrt{-1}\,H) \in 2\pi\sqrt{-1}\,\mathbb{Z}$, by Proposition 3.2.9 .

Note that since $\Delta \subset L_{\mathbb{Z}}$ we also have $\Delta \subset \hat{M}$.

Now let $\mathbb{Z}(G)^{\perp}$ be the annihilator of Z(G) in $\hat{C}$:

$$\mathbb{Z}(G) = \left\{\lambda \in \hat{C} \mid \lambda\big(\mathbb{Z}(G)\big) = 1\right\} .$$

We shall give a description of $\mathbb{Z}(G)^{\perp}$ and thereby decide when two characters of C agree on $\mathbb{Z}(G)$. Our goal is to prove

THEOREM 3.2.13 Let $\lambda = (\xi, \eta) \in \hat{C}$ (see (3.2.5) and (3.2.6)); then $\lambda \in \mathbb{Z}(G)^{\perp}$ if and only if $\eta \in L_{\mathbb{Z}}$ (see (3.2.7)).

Remark: Theorem 3.2.13 likely follows from known results about the center of G , at least when G is simply connected. The proof given here avoids the simple connectivity assumption.

Let $\mathbb{C}^{x}$ denote the non-zero complex numbers and let $\hat{\mathbb{C}^{x}}$ denote the dual group. On route toward proving theorem 3.2.13, we shall prove

THEOREM 3.2.12 Let $\ell = \dim_{\mathbb{C}} \mathfrak{h}$. Given $(\chi_1, \cdots, \chi_\ell) \in \hat{\mathbb{C}^{x}}^{\ell}$, the formula

$$(\chi_1, \cdots, \chi_\ell)(c) = \prod_{j=1}^{\ell} \chi_j\left(e^{\alpha_j(H)}\right) ,$$

$c = \exp H \in C$, $H \in \mathfrak{h}$, $\alpha_j \in \Pi$, defines an element $(\chi_1, \cdots, \chi_\ell)$ in $\mathbb{Z}(G)^\perp$. The map $(\chi_1, \cdots, \chi_\ell) \longrightarrow (\chi_1, \cdots, \chi_\ell)$ is a topological isomorphism of $\widehat{\mathbb{C}^{x^\ell}}$ onto $\mathbb{Z}(G)^\perp$.

Let $j : C \longrightarrow C/\mathbb{Z}(G)$ be the natural homomorphism. From the elementary facts about locally compact abelian groups, we know that $\mathbb{Z}(G)^\perp$ is a closed subgroup of $\hat{C}$ and the map $\gamma: \omega \rightarrow \omega \circ j$, $\omega \in \widehat{C/\mathbb{Z}(G)}$, is a topological isomorphism of $\widehat{C/\mathbb{Z}(G)}$ onto $\mathbb{Z}(G)^\perp$.

Define $\theta : C \longrightarrow \mathbb{C}^{x^\ell}$ by $\theta(c) = \left(e^{\alpha_1(H)}, \cdots, e^{\alpha_\ell(H)}\right)$ where $c = \exp H \in C$, $H \in \mathfrak{h}$, $\alpha_j \in \Pi$. Observe that the map $\theta_j : C \longrightarrow \mathbb{C}^x$ defined by $\theta_j(c) = e^{\alpha_j(H)}$ is just the (non-unitary) character (α_j, α_j) , by (3.2.5), (3.2.6) since $H = H_1 + \sqrt{-1}\, H_2 \in \mathfrak{h} = \mathfrak{a} + \sqrt{-1}\, \mathfrak{a}$ implies $\alpha_j(H) = \alpha_j(H_1) + \sqrt{-1}\, \alpha_j(H_2)$ and since $\alpha_j \in \hat{M}$, by Corollary 3.2.10. Thus θ is a well-defined, continuous homomorphism. We claim that θ is onto: Let $A = [A_{ij}]$ be the Cartan matrix:

$$A_{ij} = \frac{2(\alpha_i | \alpha_j)}{(\alpha_i | \alpha_i)} ,$$

$$1 \le i , \; j \le \ell , \quad \Pi = \{\alpha_1, \cdots, \alpha_\ell\} .$$

A is invertible; see [17]. Let $(w_1, \cdots, w_\ell) \in \mathbb{C}^{x^\ell}$. Write $w_j = e^{z_j}$, $z_j \in \mathbb{C}$. Define $a_i = \sum_{j=1}^{\ell} (A^{-1})_{ji} z_j$, $i = 1, 2, \cdots, \ell$, and set

$H = \sum_{i=1}^{\ell} a_i H_i \in \mathfrak{h}$, where $H_i = \frac{2H_{\alpha_i}}{(\alpha_i|\alpha_i)} \in \mathfrak{h}_{\mathbb{R}}$ (see (3.1.5)). For all j ,

$\alpha_j(H) = \sum_{i=1}^{\ell} a_i \alpha_j(H_i) = \sum_{i=1}^{\ell} a_i A_{ij} = \sum_{i=1}^{\ell} \sum_{k=1}^{\ell} (A^{-1})_{ki} z_k A_{ij} = \sum_{k=1}^{\ell} a_k \sum_{i=1}^{\ell} (A^{-1})_{ki} A_{ij} =$

$\sum_{k=1}^{\ell} \delta_{kj} z_k = z_j$ so $e^{\alpha_j(H)} = e^{z_j} = w_j$ implies θ is onto. Since C is σ-compact, we get $C/(\text{kernel of } \theta) \simeq \mathbb{C}^{x^{\ell}}$. However, the kernel of θ say $\mathcal{N}$, is exactly $\mathbf{Z}(G)$. For $\mathcal{N} \supset \mathbf{Z}(G)$, by Proposition 3.2.9. On the other hand suppose $c = \exp H \in \mathcal{N}$. Take $X \in \mathfrak{g}$ arbitrary; we claim that $\mathrm{Ad}(c)X = X : c \in \mathcal{N} \Longrightarrow e^{\alpha_j(H)} = 1$, $1 \le j \le \ell$, $\Longrightarrow \alpha_j(H) = 2\pi\sqrt{-1}\, n_j$, where $n_j \in \mathbb{Z}$. If $X \in \mathfrak{g}_\alpha$ for some $\alpha \in \Delta$, then, by Proposition 3.2.8, $\mathrm{Ad}(c)X = e^{\alpha(H)}X$ and since $\alpha \in L_{\mathbf{Z}}$, $\alpha_j(H) = 2\pi\sqrt{-1}\, n_j \Longrightarrow e^{\alpha(H)} = 1$. If $X \in \mathfrak{h}$, we have $\mathrm{Ad}(c)X = \mathrm{Ad}(\exp H)X = e^{\mathrm{ad}\, H}X = X$, since $\mathfrak{h}$ is abelian. Since $\mathfrak{g}$ has a root space decomposition $\mathfrak{g} = \mathfrak{h} + \sum_{\alpha \in \Delta} \mathfrak{g}_\alpha$, we get $\mathrm{Ad}(c) = $ identity on $\mathfrak{g}$ so $c \in \mathbf{Z}(G)$. Summarizing the above arguments, we get

THEOREM 3.2.11 <u>Let</u> $\Pi = \{\alpha_1, \cdots, \alpha_\ell\}$ <u>be a system of simple roots</u>. <u>Define</u> $\theta : C \longrightarrow \mathbb{C}^{x^{\ell}}$ <u>by</u> $\theta(c) = \left(e^{\alpha_1(H)}, \cdots, e^{\alpha_\ell(H)}\right)$, <u>where</u> $c = \exp H$. <u>Then</u> θ <u>is a</u> (well-defined) <u>continuous homomorphism of</u> C <u>onto</u> $\mathbb{C}^{x^{\ell}}$ <u>with kernel</u> $\mathbf{Z}(G)$. <u>Hence</u> θ <u>defines a topological isomorphism of</u> $C/\mathbf{Z}(G)$ <u>onto</u> $\mathbb{C}^{x^{\ell}}$.

θ induces a topological isomorphism $\hat{\theta}$ of $\widehat{\mathbb{C}^{\times \ell}}$ onto $\widehat{\mathbb{C}/\mathbb{Z}(G)}$:

$$\hat{\theta}(\chi)\left(j(c)\right) = \chi\left(\theta(c)\right),$$

where $(c, \chi) \in C \times \widehat{\mathbb{C}^{\times \ell}}$. On the other hand, the map $\gamma^* : \hat{\mathbb{C}}^{\times \ell} \longrightarrow \widehat{\mathbb{C}^{\times \ell}}$ defined by $\gamma^* : (\chi_1, \cdots, \chi_\ell) \longrightarrow [\chi_1, \cdots, \chi_\ell]$, where $[\chi_1, \cdots, \chi_\ell](z_1, \cdots, z_\ell) = \prod_{j=1}^{\ell} \chi_j(z_j)$ for $(\chi_1, \cdots, \chi_\ell) \in \hat{\mathbb{C}}^{\times \ell}$, $(z_1, \cdots, z_\ell) \in \mathbb{C}^{\times \ell}$, is a topological isomorphism of $\hat{\mathbb{C}}^{\times \ell}$ onto $\widehat{\mathbb{C}^{\times \ell}}$. We have

$$\hat{\mathbb{C}}^{\times \ell} \xrightarrow{\gamma^*} \widehat{\mathbb{C}^{\times \ell}} \xrightarrow{\hat{\theta}} \widehat{C/\mathbb{Z}(G)} \xrightarrow{\vee} \mathbb{Z}(G)^{\perp}$$

so the composition $\vee \circ \hat{\theta} \circ \gamma^*$ is a topological isomorphism of $\hat{\mathbb{C}}^{\times \ell}$ onto $Z(G)^{\perp}$. $\vee \circ \hat{\theta} \circ \gamma^*$ is computed immediately and, hence, we have proved Theorem 3.2.12.

Theorem 3.2.13 follows from Theorem 3.2.12. For suppose $\lambda = (\xi, \eta) \in \hat{C}$ such that λ is trivial on $\mathbb{Z}(G)$. Then by Theorem 3.2.12 there exists $(\chi_1, \cdots, \chi_\ell) \in \hat{\mathbb{C}}^{\times \ell}$ such that

$$\lambda(c) = \prod_{j=1}^{\ell} \chi_j\left(e^{\alpha_j(H)}\right) \tag{3.2.14}$$

for all $c = \exp H \in C$. Of course $\hat{\mathbb{C}}^{\times}$ is identified with $\mathbb{R} \times \mathbb{Z}$:

$$\chi \longleftrightarrow (r, m),$$

where $\chi(w) = \left(\frac{w}{|w|}\right)^m |w|^{\sqrt{-1}\, r}$ for $\chi \in \hat{\mathbb{C}}^{\times}$, $w \in \mathbb{C}^{\times}$, $(r,m) \in \mathbb{R} \times \mathbb{Z}$.

So we can choose $(r_j, m_j) \in \mathbb{R} \times \mathbb{Z}$ such that $\chi_j = (r_j, m_j), j = 1, 2, \cdots, \ell$.

Then

$$\chi_j\left(e^{\alpha_j(H)}\right) = \left(\frac{e^{\alpha_j(H)}}{\left|e^{\alpha_j(H)}\right|}\right)^{m_j} \left|e^{\alpha_j(H)}\right|^{\sqrt{-1}\, r_j}$$

$$= e^{\sqrt{-1}\, r_j \alpha_j(H_1) + \sqrt{-1}\, m_j \alpha_j(H_2)}, \tag{3.2.15}$$

where $H = H_1 + \sqrt{-1}\, H_2 \in \mathfrak{a} + \sqrt{-1}\, \mathfrak{a} = \mathfrak{h}$, since the roots are real-valued on $\mathfrak{a}$. By (3.2.5), (3.2.6), this means that

$$\chi_j = \left(r_j \sqrt{-1}\, \alpha_j, \; m_j \alpha_j\right)$$

so by (3.2.14) we get

$$\lambda = \left(\sum_{j=1}^{\ell} r_j \sqrt{-1}\, \alpha_j, \; \sum_{j=1}^{\ell} m_j \alpha_j\right),$$

i.e., $\eta = \sum_{j=1}^{\ell} m_j \alpha_j \in L_{\mathbb{Z}}$. Conversely, if $\lambda = (\xi, \eta) \in \hat{C}$, where $\eta = \sum_{j=1}^{\ell} m_j \alpha_j \in L_{\mathbb{Z}}$, then using Proposition 3.2.9, it immediately follows that λ is trivial on $\mathbb{Z}(G)$, since any $c \in \mathbb{Z}(G)$ has the form $c = \exp \sqrt{-1}\, H_2$, $H_2 \in \mathfrak{a}$; i.e., $H_1 = 0$.

Corollary 3.2.16 *Let $\lambda_1 = (\xi_1, \eta_1)$, $\lambda_2 = (\xi_2, \eta_2)$ be two characters of* C (not necessarily unitary). *Then $\lambda_1 = \lambda_2$ on $\mathbb{Z}(G)$ if and only if $\eta_1 - \eta_2 \in L_{\mathbb{Z}}$*, i.e., *if and only if η_1 and η_2 lie in the same coset of $L_{\mathbb{Z}}$ in $\hat{M}$.*

Proof: Apply Theorem 3.2.13 to the *unitary* character $(0, \eta_1 - \eta_2)$

Corollary 3.2.17 *The map*

$$\eta \longrightarrow (0, \eta)\Big|_{\mathbb{Z}(G)}$$

is a continuous homomorphism of $\hat{M}$ *onto* $\hat{\mathbb{Z}}(G)$ *with kernel* $L_{\mathbb{Z}}$. *Thus* $\hat{M}/L_{\mathbb{Z}}$ *is a finite abelian group, canonically isomorphic to* $\hat{\mathbb{Z}}(G)$.

Proof: Any unitary character of $\mathbb{Z}(G)$ extends to a unitary character of C, $\mathbb{Z}(G)$ being a closed subgroup of the locally compact abelian group C.

The fact, just remarked, that unitary characters extend will play an important role in the sequel. For the record, then, we state:

THEOREM 3.2.18 *Let* X *be a locally compact abelian group and let* X_o *be a closed subgroup of* X. *If* λ_o *is a unitary character of* X_o, *then* λ_o *extends* (non-uniquely) *to a unitary character of* X.

For a proof, see [16].

THEOREM 3.2.19 *Let* C_1 *be a closed subgroup of the Cartan subgroup* C *and let* λ_1 *be a unitary character of* C_1. *Then* λ_1 *is trivial on* $C_1 \cap \mathbb{Z}(G)$ *if and only if* λ_1 *extends to a unitary character of* C *which is trivial on* $\mathbb{Z}(G)$.

Proof: Let $j : C_1 \longrightarrow C$ be the injection map and let $\pi : C \longrightarrow C/\mathbb{Z}(G)$. Then $\pi \circ j : C_1 \longrightarrow C/\mathbb{Z}(G)$ is a continuous homomorphism and $\pi \circ j(C_1)$ is a closed subgroup of $C/\mathbb{Z}(G)$ since $\mathbb{Z}(G)C_1$ is closed in C, $\mathbb{Z}(G)$ being finite, and since

$$\pi^{-1}\left(C/\mathbb{Z}(G) - \pi \circ j(C_1)\right) = C - \mathbb{Z}(G)C_1 ,$$

which is open in C. Therefore $\pi \circ j(C_1)$ is locally compact. $\mathbb{Z}(G) \cap C_1$ is the kernel of $\pi \circ j$ and since C_1 is σ-compact, we have

$$C_1 / \left(\mathbb{Z}(G) \cap C_1\right) \simeq \pi \circ j(C_1) .$$

By duality, we have an isomorphism

$$\widehat{\pi \circ j(C_1)} \overset{\tau}{\simeq} \widehat{C_1 / \left(\mathbb{Z}(G) \cap C_1\right)} :$$

$$\tau(\chi)\left(\left(\mathbb{Z}(G) \cap C_1\right)c_1\right) = \chi\left(\pi \circ j(c_1)\right),$$

where $(\chi, c_1) \in \widehat{\pi \circ j(C_1)} \times C_1$. On the other hand, we observed earlier, there is a canonical isomorphism

$$\widehat{C_1 / \left(\mathbb{Z}(G) \cap C_1\right)} \overset{\tau^*}{\simeq} \left(\mathbb{Z}(G) \cap C_1\right)^{\perp} ,$$

$\left(\mathbb{Z}(G) \cap C_1\right)^{\perp}$ being the annihilator of $\mathbb{Z}(G) \cap C_1$ in C_1, given by

$$\tau^*(\omega) = \omega \circ \pi_1$$

where $\pi_1: C_1 \to C_1/\mathbf{Z}(G)\cap C_1$ and $\omega \in \widehat{C_1/\mathbf{Z}(G)\cap C_1}$. Therefore

$$\tau^* \circ \tau: \pi \circ j(C_1) \to (\mathbf{Z}(G) \cap C_1)^{\perp}$$

is a topological isomorphism. If $\lambda_1 \in \widehat{C}_1$ such that $\lambda_1\big|_{\mathbf{Z}(G)\cap C_1} = 1$, i.e. $\lambda_1 \in (\mathbf{Z}(G)\cap C_1)^{\perp}$, then $\lambda_1 = \tau^* \circ \tau(\chi)$ for a unique $\chi \in \widehat{\pi \circ j(C_1)}$. But since $\pi \circ j(C_1)$ is a closed subgroup of $C/\mathbf{Z}(G)$, χ extends to a unitary character χ of $C/\mathbf{Z}(G)$, by Theorem 3.2.18. Then $\chi \circ \pi \in \mathbf{Z}(G)^{\perp}$ such that for every $c_1 \in C_1$

$$(\chi \circ \pi)(c_1) = (\chi \circ \pi)(j(c_1)) = \tau(\chi)(\pi_1(c_1)) = \tau^*(\tau(\chi))(c_1) = \lambda_1(c_1)$$

so that $\chi \circ \pi\big|_{C_1} = \lambda_1$. The converse is trivial.

3.3 The induction of unitary characters of a closed subgroup of a Cartan subgroup

Let G be as in 3.1 and let $C = MA$ be a Cartan subgroup of G . Let C_1 be a closed subgroup of C . Then C_1 is a locally compact abelian group and if $\lambda_1 \in \widehat{C}_1$, the dual group of C_1, we let

$$(3.3.1) \qquad U(\cdot,\lambda_1) = \operatorname*{ind}_{C_1\uparrow G} \lambda_1 .$$

The first task is to realize $U(\cdot, \lambda_1)$ as a "multiplier" representation on an L^2 space of measurable functions. For this purpose, Theorem 3.2.18 is absolutely essential.

Since G and C_1 are unimodular we can choose a unique (up to a constant) G-invariant measure dx_1 on $C_1\backslash G$ such that

$$(3.3.2) \qquad \int_G f(x)\,dx = \int_{C_1\backslash G} \int_{C_1} f(c_1 x)\,dc_1\,d\dot{x}_1$$

for all f in $C_c(G)$, where dx, dc_1 are Haar measures on G, C_1, respectively; cf. Chapter II. Let dc, dn, dv, $d\dot{c}_1$ denote the Haar measures on the unimodular groups C, N, V, $C\backslash C_1$, respectively.

Proposition 3.3.3 <u>For all</u> k <u>in</u> $C_c(C_1\backslash G)$

$$\int_{C_1\backslash G} k(C_1 x)\,d\dot{x}_1 = \int_N \int_V \int_{C_1\backslash C} k(C_1 cnv)\,d\dot{c}_1\,dv\,dn.$$

<u>Proof:</u> Given k in $C_c(C_1\backslash G)$ we can always choose a ϕ in $C_c(G)$ so that

$$k(C_1 x) = \int_{C_1} \phi(c_1 x)\,dc_1$$

for all x in G. By (3.3.2) and

$$(3.1.24) \qquad \int_{C_1\backslash G} k(C_1 x)\,d\dot{x}_1 = \int_G \phi(x)\,dx = \int_N \int_V \int_C \phi(cnv)\,dc\,dv\,dn\,.$$

Analogous to (3.3.2), we have

$$\int_C \psi(c)dc = \int_{C_1\backslash C} \int_{C_1} \psi(c_1 c)dc_1 \, d\bar{c}_1$$

for any ψ in $C_c(C)$. Hence

$$\int_C \phi(cnv)dc = \int_{C_1\backslash C} \int_{C_1} \phi(c_1 cnv)dc_1 \, d\bar{c}_1$$

$$= \int_{C_1\backslash C} k(C_1 cnv)d\bar{c}_1$$

for all $(n, v) \in N \times V$.

Now consider the induced representation $U(\cdot, \lambda_1) = \operatorname{ind}_{C_1 \uparrow G} \lambda_1$, for $\lambda_1 \in \hat{C}_1$. In accordance with 2.2, $U(\cdot, \lambda_1)$ acts on the space $H(\lambda_1)$ of measurable functions

$$f : G \longrightarrow \mathbb{C}$$

such that

(i) $$f(c_1 x) = \lambda_1(c_1)\, f(x)$$

for all $(c_1, x) \in C_1 \times G$

(ii) $$\int_{C_1\backslash G} |f(x)|^2 \, d\bar{x}_1 < \infty .$$

The action of $U(a, \lambda_1)$ on a function f in $H(\lambda_1)$, $a \in G$, is right-translation of f by a.

Proposition 3.3.4 <u>Let</u> λ_1 <u>be an extension of</u> λ_1 <u>to a unitary character of</u> C. <u>Then the formula</u>

$$(\Phi f)(n, v, C_1 c) = \overline{\lambda}_1 (c)\, f(cnv),$$

<u>for</u> $f \in H(\lambda_1)$, $(c, n, v) \in C \times N \times V$, <u>defines a unitary map</u> Φ <u>of</u> $H(\lambda_1)$ <u>onto</u> $L^2(N \times V \times C_1 \backslash C)$.

<u>Proof</u>: Φ is well-defined for suppose $C_1 c = C_1 d$, $c, d \in C$. Then $c_1 c = d$ for some $c_1 \in C_1$ and since $f \in H(\lambda_1)$, we have

$$\overline{\lambda}_1(d)\, f(dnv) = \overline{\lambda}_1(c_1)\, \overline{\lambda}_1(c)\, \lambda_1(c)\, f(cnv) = \overline{\lambda}_1(c)\, f(cnv)$$

for all $(n, v) \in N \times V$. By Proposition 3.3.3

$$\int_N \int_V \int_{C_1 \backslash C} |\Phi f(n, v, C_1 c)|^2\, d\overline{c}_1\, dv\, dn =$$

$$\int_N \int_V \int_{C_1 \backslash C} |f(cnv)|^2\, d\overline{c}_1\, dv\, dn = \int_{C_1 \backslash G} |f(x)|^2\, d\overline{x}_1,$$

so Φ preserves norms. Φ^{-1} is given by

$$(\Phi^{-1} g)(x) = \begin{cases} \lambda_1(c)\, g(n, v, C_1 c) & \text{if } x = cnv \in CNV \\ 0 & \text{if } x \in G - CNV \end{cases},$$

$g \in L^2(N \times V \times C_1 \backslash C)$; recall from <u>3.1</u> that CNV is dense in G. Thus Φ is unitary.

It follows that $U(\cdot, \lambda)$ can be realized on the Hilbert space $L^2(N \times V \times C_1 \backslash C)$: Let $g = \Phi f \in L^2(N \times V \times C_1 \backslash C)$, where $f \in H(\lambda_1)$. Then for $a \in G$ and $(n, v, C_1 c) \in N \times V \times C_1 \backslash C$

$$\left(\Phi\, U(a, \lambda_1)\, \Phi^{-1} g\right)(n, v, C_1 c) = \bar{\lambda}_1(c)\left(U(a, \lambda_1) f\right)(cnv)$$

$$= \bar{\lambda}_1(c)\, f(cnva)$$

$$= \bar{\lambda}_1(c)\, f\left(c\, c(va)\, c(va)^{-1}\, n\, c(va)\, n(va)\, v(va)\right),$$

where we write $va = c(va)\, n(va)\, v(va) \in CNV$. Now C normalizes N (and V) so $c(va)^{-1}\, n\, c(va) \in N$. Also we observed that $(\Phi^{-1} g)(c_1 n_1 v_1) = \lambda_1(c_1)\, g(n_1, v_1, C_1 c_1)$ for $(c_1, n_1, v_1) \in C \times N \times V$. Since $f = \Phi^{-1} g$ we get

$$\bar{\lambda}_1(c)\, f\left(c\, c(va)\, c(va)^{-1} n\, c(va)\, n(va)\, v(va)\right)$$

$$= \bar{\lambda}_1(c)\, \lambda_1\left(c\, c(va)\right) g\left(c(va)^{-1} n\, c(va)\, n(va),\ v(va), C_1\, c\, c(va)\right).$$

But $\bar{\lambda}_1(c)\, \lambda_1\left(c\, c(va)\right) = \lambda_1\left(c(va)\right)$. This gives

THEOREM 3.3.5 Let G be a connected complex semisimple Lie group and let C be a Cartan subgroup of G. If C_1 is a closed subgroup of C and if λ_1 is a unitary character of C_1, then the induced representation $U(\cdot, \lambda_1) = \operatorname{ind}_{C_1 \uparrow G} \lambda_1$ can be realized on the Hilbert space $L^2(N \times V \times C_1\backslash C)$ as follows:

$$\left(U(a, \lambda_1) f\right)(n, v, C_1 c) = \lambda_1\left(c(va)\right) f\left(c(va)^{-1} n\, c(va)\, n(va),\ v(va),\ C_1 c\, c(va)\right),$$

where $f \in L^2(N \times V \times C_1\backslash C)$, $a \in G$, $(n, v, C_1 c) \in N \times V \times C_1\backslash C$, and λ_1 is any unitary character of C which extends λ_1; N and V are defined in 3.1.

Remark: Theorem 3.3.5 can be obtained by inducing in stages. One induces from C_1 to C and then from C to G.

We shall be interested in taking a closer look at the family of representations $\operatorname{ind}_{C_1 \uparrow G} \lambda_1$, where λ_1 varies over $\hat{C}_1$. It will be important, for example, to know when two such representations are unitarily equivalent.

Proposition 3.3.6 Let $\lambda_1, \gamma_1 \in \hat{C}_1$. For every $z \in \mathbb{Z}(G) \cap C_1$

$$U(z, \lambda_1) = \lambda_1(z)\, I,$$

where I is the identity operator. In particular if

$$U(\cdot, \lambda_1) \simeq U(\cdot, \gamma_1),$$

then we must have $\gamma_1 = \lambda_1$ on $\mathbb{Z}(G) \cap C_1$.

Proof: Let $(c, n, v) \in C \times N \times V$ be arbitrary, $z \in \mathbb{Z}(G) \cap C_1$. Then

$$vz = zv$$

and since $\mathbb{Z}(G) \subset M \subset C$, this means that

$$c(vz) = z, \quad n(vz) = e, \quad v(vz) = v.$$

Also

$$c(vz)^{-1} n\, c(vz)\, n(vz) = z^{-1} n\, z\, e = n.$$

Since $z \in C_1$ we get $C_1 cz = C_1 c$. Proposition 3.3.6 now follows immediately from Theorem 3.3.5.

3.4 A preliminary theorem

We shall devote this section to proving the following fundamental theorem:

THEOREM 3.4.1 Let G be a complex connected semi-simple Lie Group. Let C be a Cartan subgroup of G and let C_1 be a closed subgroup of C . Given a unitary character λ_1 of C_1 denote the unitary representation of G induced by λ_1 by $U(\cdot,\lambda_1)$:

$$U(\cdot,\lambda_1) = \operatorname*{ind}_{C_1 \uparrow G} \lambda_1 .$$

Suppose α is any simple root and $(s,m) \in \sqrt{-1}\,\mathbb{R} \times \mathbb{Z}$ is arbitrary. Let $\lambda_{(\alpha,s,m)}$ be the unitary character

$$\lambda_{(\alpha,s,m)} = (s\alpha, m\alpha)$$

of C (see (3.2.5)). Then

$$U(\cdot,\lambda_1) \simeq U(\cdot,\lambda_1\, \lambda_{(\alpha,s,m)}\big|_{C_1}) .$$

In the proof the unitary equivalence of the representations $U(\cdot,\lambda_1)$, $U(\cdot,\lambda_1\, \lambda_{(\alpha,s,m)}\big|_{C_1})$ is exhibited explicitly. We begin by considering the multiplier realization of $U(\cdot, \lambda)$ on $L^2(N\times V\times C_1\backslash C)$ given in Theorem 3.3.5:

$$(3.4.2) \qquad \big(U(a,\lambda_1)\, f\big)(n, v, C_1\, c)$$

$$= \lambda_1\big(c(va)\big)\, f\big(c(va)^{-1} n\, c(va)\, n(va)\ ,\ v(va)\ ,\ C_1\, c\, c(va)\big)$$

where λ_1 is any unitary character of C which extends λ_1, $a \in G$, $f \in L^2(N\times V\times C_1\backslash C)$. Given the simple root α, let p_α be the corresponding simple Weyl reflection (see (3.1.14)). By Theorem 3.1.19 and the remarks following Theorem 3.1.19 we can write

$$(3.4.3) \qquad N = N_{p_\alpha} N'_{p_\alpha}$$

where $N_{p_\alpha} \cap N'_{p_\alpha} = \{e\}$, and

$$(3.4.4) \qquad z \longrightarrow \exp z X_\alpha$$

for non-zero $X_\alpha \in \mathfrak{g}_\alpha$, is an analytic isomorphism of $\mathbb{C}$ onto N_{p_α}. We shall denote N_{p_α}, N'_{p_α} by N_α, N'_α, and $L^2(N\times V\times C_1\backslash C)$, $L^2(C\times N'_\alpha\times V\times C_1\backslash C)$ by $\mathcal{H}$, $\mathcal{H}_\alpha$, respectively, in order to simplify the notation.

Since N is simply connected, we can use the exponential map

$$\exp : \mathfrak{n} \longrightarrow N$$

to identify Haar measure on N with Lebesgue measure on Euclidean space. Then (3.4.3) and (3.4.4) mean that we can define a unitary map

$$\Phi_\alpha : \mathcal{H} \longrightarrow \mathcal{K}_\alpha$$

by the formula

$$(3.4.5) \qquad (\Phi_\alpha f)(z, n', v, C_1 c) = f\big((\exp z X_\alpha) n', v, C_1 c\big),$$

where $f \in \mathcal{H}$, $n' \in N'_\alpha$, $v \in V$, $c \in C$, $z \in \mathbb{C}$. To compute the transform of $U(\cdot, \lambda_1)$ by Φ_α, i.e., the representation $\Phi_\alpha U(\cdot, \lambda_1) \Phi_\alpha^{-1}$ on $\mathcal{K}_\alpha$, we use the familiar rule:

$$(3.4.6) \qquad a \exp X a^{-1} = \exp \mathrm{Ad}(a) X$$

for any $a \in G$, $X \in \mathfrak{g}$ ($\mathfrak{g}$ is the Lie algebra of G).

We have, for $g \in \mathcal{K}$

$$\big(\Phi_\alpha U(a, \lambda_1) \Phi_\alpha^{-1} g\big)(z, n', v, C_1 c)$$

$$= \big(U(a, \lambda_1) f\big)\big((\exp z X_\alpha) n', v, C_1 c\big),$$

where $\Phi_\alpha f = g$,

$$= \lambda_1\big(c(va)\big) f\big(c(va)^{-1} (\exp z X_\alpha) n' c(va)\, n(va), v(va), C_1 c\, c(va)\big),$$

by (3.4.2),

$$= \lambda_1\Big(c(va)\Big)\, f\Big(c(va)^{-1} \exp z\, X_\alpha\, c(va)\, n_2\, ,\ v(va)\, ,\ C_1\, c\, c(va)\Big),$$

where $$n_2 = c(va)^{-1}\, n^1\, c(va)\, n(va)\, .$$

By (3.4.6)

$$c(va)^{-1} \exp z\, X_\alpha\, c(va) = \exp \mathrm{Ad}\Big(c(va)^{-1}\Big)\, z\, X_\alpha\, .$$

Write $$c(va)^{-1} = \exp H_{va}\, ,$$

where $H_{va} \in \mathfrak{h}$ ($\mathfrak{h}$ is the Lie algebra of C). Then by Proposition 3.2.8 ,

$$\mathrm{Ad}\Big(c(va)^{-1}\Big)\, z\, X_\alpha = e^{\alpha(H_{va})}\, z\, X_\alpha\, .$$

Therefore

$$c(va)^{-1} \exp z\, X_\alpha\, c(va) = \exp e^{\alpha(H_{va})}\, z\, X_\alpha$$

and since $\Phi_\alpha f = g$, we get

Proposition 3.4.7 <u>On the space</u> $\mathcal{K}_\alpha$, <u>the representation</u> $U(\cdot, \lambda_1)$ is given by

$$\Big(U(a, \lambda_1)\, g\Big)(z, n', v, C_1\, c)$$

$$= \lambda_1\Big(c(va)\Big)\, g\Big(e^{\alpha(H_{va})} z\, ,\ c(va)^{-1}\, n'\, c(va)\, n(va)\, ,\ v(va), C_1\, c\, c(va)\Big),$$

where $$c(va)^{-1} = \exp H_{va}\, .$$

Next we transform the representation $U(\cdot, \lambda_1)$ on $\mathcal{H}_\alpha$ via the Fourier transform, F, on $\mathbb{C}$. By the Plancherel Theorem for $\mathbb{C}$, F is a unitary mapping of $L^2(\mathbb{C})$ such that for $\phi \in L^1(\mathbb{C}) \cap L^2(\mathbb{C})$,

$$(F\phi)(w) = \frac{1}{2\pi}\int_{\mathbb{C}} e^{-\sqrt{-1}\,\mathrm{Re}\, z\overline{w}}\,\phi(z)\,dz .$$

Let

$$F_\alpha = F \otimes I \otimes I \otimes I, \qquad \text{i.e.,}$$

(3.4.8) $$(F_\alpha \tilde{\phi})(w, n', v, C_1 c) = \frac{1}{2\pi}\int_{\mathbb{C}} e^{-\sqrt{-1}\,\mathrm{Re}\, z\overline{w}}\,\tilde{\phi}(z, n', v, C_1 c)\,dz,$$

for $\tilde{\phi}$ a nice function on $\mathbb{C} \times N'_\alpha \times V \times C_1\backslash C$. Equation (3.4.8) defines a unitary map F_α of $\mathcal{H}_\alpha$ and we want to compute $F_\alpha U(\cdot, \lambda_1) F_\alpha^{-1}$.

For $g \in \mathcal{H}_\alpha$ we have

$$\left(F_\alpha\, U(a, \lambda_1)\, F_\alpha^{-1} g\right)(w, n', v, C_1 c)$$

$$= \frac{1}{2\pi}\int_{\mathbb{C}} e^{-\sqrt{-1}\,\mathrm{Re}\, z\overline{w}}\left(U(a, \lambda_1)\, f\right)(z, n', v, C_1 d)\,dz ,$$

where

$$F_\alpha f = g ,$$

$$= \frac{1}{2\pi}\int_{\mathbb{C}} e^{-\sqrt{-1}\,\mathrm{Re}\, z\overline{w}}\,\lambda_1\left(c(va)\right) f\left(e^{\alpha(H_{va})} z,\ n_2,\ v(va),\ C_1 c\, c(va)\right) dz ,$$

by Proposition 3.4.7, where

$$n_2 = c(va)^{-1}\, n'\, c(va)\, n(va)\,, \quad c(va)^{-1} = \exp H_{va}\,.$$

Now make the change of variables

$$z \longrightarrow e^{-\alpha(H_{va})} z\,.$$

The Jacobian is $\left| e^{-\alpha(H_{va})} \right|^2 = e^{-2\,\mathrm{Re}\,\alpha(H_{va})}$ so that

$$\int_{\mathbb{C}} e^{-\sqrt{-1}\,\mathrm{Re}\, z\overline{w}} f\left(e^{\alpha(H_{va})} z,\, n_2,\, v(va),\, C_1 c\, c(va)\right) dz$$

$$= e^{-2\,\mathrm{Re}\,\alpha(H_{va})} \int_{\mathbb{C}} e^{-\sqrt{-1}\,\mathrm{Re}\, z\, \overline{e^{-\alpha(H_{va})} w}} f\left(z,\, n_2,\, v(va),\, C_1 c\, c(va)\right) dz$$

$$= e^{-2\,\mathrm{Re}\,\alpha(H_{va})}\, 2\pi\, (Ff)\left(e^{-\overline{\alpha(H_{va})}} w,\, n_2,\, v(va),\, C_1 c\, c(va)\right).$$

Since $F_\alpha f = g$, we get

Proposition 3.4.9 <u>The representation</u> $U(\cdot, \lambda_1)$ <u>is given on the space</u> $\mathcal{K}_\alpha$ <u>by</u>

$$\left(U(a, \lambda_1)\, g\right)(w, n', v, C_1 c)$$

$$= e^{-2\,\mathrm{Re}\,\alpha(H_{va})}\, \lambda_1\left(c(va)\right) g\left(e^{-\overline{\alpha(H_{va})}} w,\, n_2,\, v(va),\, C_1 c\, c(va)\right)$$

where
$$n_2 = c(va)^{-1}\, n'\, c(va)\, n(va)$$

$$c(va)^{-1} = \exp H_{va}\,.$$

If w is a non-zero complex number, we define

$$[w] = \frac{w}{|w|}.$$

Suppose $(s, m) \in \sqrt{-1}\,\mathbb{R} \times \mathbb{Z}$. Define

$$\lambda_{(\alpha, s, m)} = (s\alpha, m\alpha).$$

Then $\lambda_{(\alpha, s, m)} \in \hat{C}$ and for $c = \exp H \in C$, $H = H_1 + \sqrt{-1}\, H_2 \in \mathfrak{a} + \sqrt{-1}\,\mathfrak{a} = \mathfrak{h}$, we have

$$\lambda_{(\alpha, s, m)}(c) = e^{s\alpha(H_1) + \sqrt{-1}\, m\alpha(H_2)}. \tag{3.4.10}$$

We construct a unitary map

$$B = B(s, m)$$

of $\mathcal{K}_\alpha$ by setting

$$(Bg)(w, n', v, C_1 c) = |w|^{-s} [w]^m g(w, n', v, C_1 c).$$

Using Proposition 3.4.9 we have, for $f = B^{-1} g$

$$\left(B\, U(a, \lambda_1)\, B^{-1} g\right)(w, n', v, C_1 c)$$

$$= |w|^{-s} [w]^m \left(U(a, \lambda_1) f\right)(w, n', v, C_1 c)$$

$$= |w|^{-s} [w]^m e^{-2 \operatorname{Re} \alpha(H_{va})} \lambda_1\left(c(va)\right) f\left(e^{-\overline{\alpha(H_{va})}} w, n_2, v(va), C_1 c\, c(va)\right),$$

where $n_2 = c(va)^{-1}\, n'\, c(va)\, n(va)$, $c(va)^{-1} = \exp H_{va}$. Write

$$|w|^{-s}\,[w]^m$$

$$= |w|^{-s}[w]^m \left| e^{-\overline{\alpha(H_{va})}} w \right|^s \left[e^{-\overline{\alpha(H_{va})}} w \right]^{-m} \left| e^{-\overline{\alpha(H_{va})}} w \right|^{-s} \left[e^{-\overline{\alpha(H_{va})}} w \right]^m .$$

Now

$$|w|^{-s}\,[w]^m \left| e^{-\overline{\alpha(H_{va})}} w \right|^s \left[e^{-\alpha(H_{va})} w \right]^{-m}$$

$$= e^{-s\,\mathrm{Re}\,\alpha(H_{va}) - m\sqrt{-1}\,\mathrm{Im}\,\alpha(H_{va})}$$

$$= \lambda_{(\alpha,\, s,\, m)} \left(c(va) \right),$$

by (3.2.10). For

$$c(va)^{-1} = \exp H_{va}$$

implies that

$$c(va) = \exp(-H_{va})$$

and the roots are real-valued on $\mathfrak{a}$. Moreover $Bf = g$, so we get

$$\left(B\, U(a, \lambda_1)\, B^{-1} g \right)(w, n', v, C_1 c)$$

$$= \lambda_1\, \lambda_{(\alpha,s,m)}\left(c(va)\right) e^{-2\,\mathrm{Re}\,\alpha(H_{va})}\, g\left(e^{-\overline{\alpha(H_{va})}} w,\, n_2,\, v(va),\, C_1 c\; c(va) \right) =$$

$$= \left(U\left(a, \lambda_1 \lambda_{(\alpha, s, m)}\Big|_{C_1}\right) g\right)(w, n', v, C_1 c),$$

by Proposition 3.4.9. Therefore, on the space $\mathcal{K}_\alpha$, the operator $B(s, m)$ defines a unitary equivalence of $U(\cdot, \lambda_1)$ and $U\left(\cdot\ \lambda_1 \lambda_{(\alpha, s, m)}\Big|_{C_1}\right)$. This completes the proof of Theorem 3.4.1.

Consider the diagram:

$$\mathcal{H} \xrightarrow{\Phi_\alpha} \mathcal{K}_\alpha \xrightarrow{F_\alpha} \mathcal{K}_\alpha \xrightarrow{B(s,m)} \mathcal{K}_\alpha \xrightarrow{F_\alpha^{-1}} \mathcal{K}_\alpha \xrightarrow{\Phi_\alpha^{-1}} \mathcal{H}.$$

If we set

$$A(\alpha, s, m) = F_\alpha^{-1} B(s, m) F_\alpha, \tag{3.4.11}$$

$$\psi(\alpha, s, m) = \Phi_\alpha^{-1} A(\alpha, s, m) \Phi_\alpha, \tag{3.4.12}$$

then $\psi(\alpha, s, m)$ is a unitary map of

$$\mathcal{H} = L^2(N \times V \times C\backslash C_1)$$

such that

$$\psi(\alpha, s, m)\, U(\cdot, \lambda_1)\, \psi(\alpha, s, m)^{-1} = U\left(\cdot, \lambda_1 \lambda_{(\alpha, s, m)}\Big|_{C_1}\right). \tag{3.4.13}$$

The operator $A(\alpha, s, m)$ on $\mathcal{K}_\alpha$ can be described as follows:

Formally

$$\left(A(\alpha, s, m)\, f\right)(z, n, v, C_1 c) = \frac{1}{\gamma(s, m)} \int_{\mathbb{C}} |t|^{s-2} [t]^m\, f(z+t, n, v, C_1 c)\, dt ,$$

where

$$\gamma(s, m) = \frac{2^s\, \pi\, \sqrt{-1}^{\,m}\, \Gamma\left(\frac{|m| + s}{2}\right)}{\Gamma\left(\frac{|m| - s + 2}{2}\right)} .$$

To see this, one simply imitates the arguments given by Kunze and Stein in [19]. Their arguments, interpreted and applied in the present context, give the following

Proposition 3.4.14 Let $(m, s') \in \mathbb{Z} \times \mathbb{C}$ such that $0 < \mathrm{Re}(s') < 1$. If f is in $C_c(\mathbb{C} \times N'_\alpha \times V \times C_1 \backslash C)$, then

(i) the integral

$$\frac{1}{\gamma(s', m)} \int_{\mathbb{C}} |t|^{s'-2} [t]^m\, f(z+t, n, v, C_1 c)\, dt$$

converges for all $(z, n, v, c) \in \mathbb{C} \times N \times V \times C$ and defines a continuous function $\tilde{A}(\alpha, s', m) f$ in $L^2(\mathbb{C} \times N'_\alpha \times V \times C_1/C)$

(ii) the mapping $s' \longrightarrow \tilde{A}(\alpha, s', m) f$ has a unique extension which is continuous in the strip $0 \le \mathrm{Re}(s') < 1$, as a function of s' with values in $L^2(\mathbb{C} \times N'_\alpha \times V \times C_1/C)$

(iii) <u>if</u> $\mathrm{Re}(s') = 0$, <u>the operator</u> $\widetilde{A}(\alpha, s'm)$: $f \to \widetilde{A}(\alpha, s', m)f$ <u>is isometric on</u> $C_c(\mathbb{C} \times N'_\alpha \times V \times C/C_1)$ <u>and extends uniquely to a unitary operator on</u> $L^2(\mathbb{C} \times N'_\alpha \times V \times C/C_1)$.

We have then

$$A(\alpha, s, m)f = \lim_{\mathrm{Re}(s')\to 0} \widetilde{A}(\alpha, s', m)f ,$$

where the limit is taken in L^2. Considering (3.4.12), we define $\psi(\alpha, s, m)$ <u>formally</u> on $L^2(N \times V \times C/C_1)$ by:

$$(\psi(\alpha, s, m)f)(n, v, C_1c) = \frac{1}{\gamma(s, m)} \int_{\mathbb{C}} |t|^{s-2}[t]^m f((\exp t X_\alpha)n, v, C_1c)dt .$$

$\psi(\alpha, s, m)f)$ is given, rigorously, as a limit of the right-hand side, in L^2:

$$\lim_{\mathrm{Re}(s')\to 0} \frac{1}{\gamma(s'm)} \int_{\mathbb{C}} |t|^{s'-2}[t]^m f((\exp t X_\alpha)n, v, C_1c)dt .$$

<u>3.5</u> The relative residue of a character and the unitary equivalence of the representations $\operatorname{ind}_{C_1\uparrow G} \lambda_1$, $\lambda_1 \in \hat{C}_1$

Proposition 3.5.1 <u>The map</u> $\eta \to (0, \eta)|_{C_1 \cap \mathbf{Z}(G)}$ <u>is a continuous homomorphism of</u> $\widehat{M}$ <u>onto</u> $C_1 \cap \mathbf{Z}(G)$ <u>with kernel</u>

$$\eta_{C_1} = \{\eta \in \hat{M} \mid (0, \eta - \eta_o) \in C_1^{\perp} \ \underline{\text{for some}} \ \eta_o \in L_{\mathbb{Z}}\} ;$$

hence $\hat{M}/\eta_{C_1} \simeq \widehat{C_1 \cap \mathbb{Z}(G)}$.

Proof: Let $\chi \in \widehat{C_1 \cap \mathbb{Z}(G)}$. By Theorem 3.2.18, χ extends to a unitary character $\tilde{\chi}$ of $\mathbb{Z}(G)$. By Corollary 3.2.17 $\tilde{\chi} = (0, \eta)$ for some $\eta \in \hat{M}$ so that $(0, \eta)|_{C_1 \cap \mathbb{Z}(G)} = \chi$. Hence $\hat{M}$ maps onto $\widehat{C_1 \cap \mathbb{Z}(G)}$. If $\eta \in \eta_{C_1}$, then $(0, \eta)|_{C_1} \in \hat{C_1}$ and $(0, \eta)|_{C_1}$ is trivial on $C_1 \cap \mathbb{Z}(G)$. Then by Theorem 3.2.19 $(0, \eta)|_{C_1}$ extends to a unitary character $(\tilde{\xi}, \tilde{\eta})$ of C which is trivial on $\mathbb{Z}(G)$. From Theorem 3.2.13 $\tilde{\eta} \in L_{\mathbb{Z}}$. Clearly $(0, \tilde{\eta})$ extends $(0, \eta)|_{C_1}$ as well, i.e., $(0, \eta - \tilde{\eta}) \in C_1^{\perp}$. Conversely, if $\eta \in \hat{M}$ is such that $(0, \eta - \tilde{\eta}) \in C_1^{\perp}$ for some $\tilde{\eta} \in L_{\mathbb{Z}}$, then by Theorem 3.2.13, again, $(0, \eta) = (0, \tilde{\eta})(0, \eta - \tilde{\eta})$ is trivial on $C_1 \cap \mathbb{Z}(G)$.

Remark: If $C_1 = C$, then $\eta_{C_1} = L_{\mathbb{Z}}$ so that Proposition 3.5.1 reduces to Corollary 3.2.17.

Definition 3.5.2 Let C_1 be a closed subgroup of the Cartan subgroup C. Choose once and for all a set of representatives

$\rho_1, \cdots, \rho_{n_1} \in \hat{M}$ for the cosets of $\mathcal{N}_{C_1}$ in $\hat{M}$; n_1 = order of $C_1 \cap \mathbb{Z}(G)$. If $\lambda = (\xi, \eta)$ is any character of C (λ not necessarily unitary), the residue of λ relative to C_1 (or the relative residue of λ) is the unitary character $(0, \rho_j)$ of C, where j is the index such that $\eta - \rho_j \in \mathcal{N}_{C_1}$. We shall also call ρ_j the relative residue of η and write $\rho_j = \rho_\eta$. If $C_1 = C$, $(0, \rho_j)(\rho_j)$ is simply called the residue of $\lambda(\eta)$.

Proposition 3.5.3 If $\lambda_j = (\xi_j, \eta_j)$, $j = 1, 2$ are characters of C, then $\lambda_1 = \lambda_2$ on $C_1 \cap \mathbb{Z}(G)$ if and only if $\eta_1 - \eta_2 \in \mathcal{N}_{C_1}$. Equivalently $\lambda_1 = \lambda_2$ on $C_1 \cap \mathbb{Z}(G)$ if and only if λ_1 and λ_2 have the same relative residue.

Proof: If $\eta_1 - \eta_2 = n \in \mathcal{N}_{C_1}$, then for all $c \in C_1 \cap \mathbb{Z}(G)$, $\lambda_1(c) = (\xi_1, \eta_1)(c) = (\xi_1, \eta_2)(c)(0, n)(c) = (\xi_1, \eta_2)(c)$ (by Proposition 3.5.1) $= (0, \eta_2)(c)$ (since $c \in M$) $= (\xi_2, \eta_2)(c)$ (since $c \in M$) $= \lambda_2(c)$. Conversely suppose $\lambda_1 = \lambda_2$ on $C_1 \cap \mathbb{Z}(G)$. Then

$$(0, \eta_1 - \eta_2)\Big|_{C_1 \cap \mathbb{Z}(G)} = (\xi_1 - \xi_2, \eta_1 - \eta_2\Big|_{C_1 \cap \mathbb{Z}(G)} = 1$$

so that $\eta_1 - \eta_2 \in \mathcal{N}_{C_1}$.

Proposition 3.5.4 Let $\gamma, \lambda \in \hat{C}_1$ be such that $\gamma = \lambda$ on $C_1 \cap \mathbb{Z}(G)$. Then we can choose extensions $\tilde{\gamma}, \tilde{\lambda}$ of γ, λ, respectively, in $\hat{C}$ such that $\tilde{\gamma} = \tilde{\lambda}$ on $\mathbb{Z}(G)$.

Proof: Let $\tilde{\lambda}$ be any unitary character of C which extends λ. Since $\gamma\lambda^{-1} = 1$ on $C_1 \cap \mathbb{Z}(G)$, $\gamma\lambda^{-1}$ extends to a unitary character β of C such that $\beta = 1$ on $\mathbb{Z}(G)$; this follows by Theorem 3.2.19. Put $\tilde{\gamma} = \tilde{\lambda}\beta \in \hat{C}$. Then $\tilde{\gamma}\big|_{C_1} = \gamma$ and $\tilde{\gamma} = \tilde{\lambda}$ on $\mathbb{Z}(G)$.

The next theorem gives the main result of this chapter.

THEOREM 3.5.5 *Let G be a complex connected semi-simple Lie group. Let C be a Cartan subgroup of G and let C_1 be a closed subgroup of C. If $\lambda \in \hat{C}_1$ and if $\tilde{\lambda} = (\xi, \eta) \in \hat{C}$ is any extension of λ to C, then*

$$\operatorname{ind}_{C_1 \uparrow G} \lambda \simeq \operatorname{ind}_{C_1 \uparrow G} (0, \rho_\eta)\big|_{C_1},$$

where ρ_η is the relative residue of η. If $\lambda, \gamma \in \hat{C}_1$, then

$$\operatorname{ind}_{C_1 \uparrow G} \lambda \simeq \operatorname{ind}_{C_1 \uparrow G} \gamma$$

if and only if $\lambda = \gamma$ on $C_1 \cap \mathbb{Z}(G)$. The family of representations $\{\operatorname{ind}_{C_1 \uparrow G} \lambda \mid \lambda \in \hat{C}_1\}$ breaks up into exactly n_1 unitary equivalence classes, where n_1 is the order of $C_1 \cap \mathbb{Z}(G)$. In particular, if $C_1 = C$, then, within equivalence, $\operatorname{ind}_{C \uparrow G} (\xi_1, \eta_1)$ depends only on the coset of $L_{\mathbb{Z}}$ in $\hat{M}$ which contains η_1, for any $(\xi_1, \eta_1) \in \hat{C}$.

<u>Proof</u>: The main idea is to apply Theorem 3.4.1 finitely many times. Let $\lambda \in \hat{C}_1$ and extend λ to a unitary character $\tilde{\lambda} = (\xi, \eta)$ of C. Since the simple roots $\Pi = \{\alpha_1, \cdots, \alpha_\ell\}$ form an $\mathbb{R}$ basis for $h^*_{\mathbb{R}}$, we can write $\xi = \sum_{j=1}^{\ell} s_j \alpha_j$, $\eta = \rho_\eta + n$, where each $s_j \in \sqrt{-1}\,\mathbb{R}$, ρ_η is the relative residue of η, and $n \in \mathfrak{n}_{C_1}$. By Proposition 3.5.1 there exist integers m_j such that $\left(0, n - \sum_{j=1}^{\ell} m_j \alpha_j\right) \in C_1^{\perp}$. Define $\lambda_j = (-s_j \alpha_j, -m_j \alpha_j) \in \hat{C}$. Then by Theorem 3.4.1

$$U(\cdot, \lambda) \simeq U\left(\cdot, \lambda \lambda_1\Big|_{C_1}\right) \simeq U\left(\cdot, \lambda \lambda_1\Big|_{C_1} \lambda_2\Big|_{C_1}\right) \simeq \cdots \simeq U\left(\cdot, \lambda \prod_{j=1}^{\ell} \lambda_j\Big|_{C_1}\right).$$

Now $\lambda \prod_{j=1}^{\ell} \lambda_j\Big|_{C_1} = \left(\tilde{\lambda} \prod_{j=1}^{\ell} \lambda_j\right)\Big|_{C_1} = \left(0, \eta - \sum_{j=1}^{\ell} m_j \alpha_j\right)\Big|_{C_1} =$

$\left(0, \rho_\eta + n - \sum_{j=1}^{\ell} m_j \alpha_j\right)\Big|_{C_1} = (0, \rho_\eta)\Big|_{C_1} \cdot \left(0, n - \sum_{j=1}^{\ell} m_j \alpha_j\right)\Big|_{C_1} =$

$(0, \rho_\eta)\Big|_{C_1}$, since $\left(0, n - \sum_{j=1}^{\ell} m_j \alpha_j\right) \in C_1^{\perp}$. Therefore $U(\cdot, \lambda) \simeq U\left(\cdot, (0, \rho_\eta)\Big|_{C_1}\right)$. Next take $\lambda, \gamma \in \hat{C}_1$. If $U(\cdot, \lambda) \simeq U(\cdot, \gamma)$, then $\lambda = \gamma$ on $C_1 \cap \mathbb{Z}(G)$ by Proposition 3.3.6. Conversely suppose $\lambda = \gamma$ on $C_1 \cap \mathbb{Z}(G)$. By Proposition 3.5.4, we can choose extensions $\tilde{\lambda}, \tilde{\gamma}$ in $\hat{C}$ such that $\tilde{\lambda} = \tilde{\gamma}$ on $\mathbb{Z}(G)$. Let $\tilde{\lambda} = (\xi_1, \eta_1)$ and $\tilde{\gamma} = (\xi_2, \eta_2)$. Then, since $\tilde{\lambda} = \tilde{\gamma}$ on $C_1 \cap \mathbb{Z}(G)$, we have $\rho_{\eta_1} = \rho_{\eta_2}$ by Proposition 3.5.3. By what we have just shown above, we get

$U(\cdot,\lambda) \simeq U(\cdot(0,\rho_{\eta_1})\big|_{C_1}) = U(\cdot(0,\rho_{\eta_2})\big|_{C_1}) \simeq U(\cdot,\gamma)$. Suppose the unitary characters of $C_1 \cap \mathbf{Z}(G)$ are $\chi_1,\ldots,\chi_{n_1}$. Extend each χ_j to a unitary character $\widetilde{\chi}_j$ of C_1 . If $\gamma \in \hat{C}_1$ then $\lambda|_{C_1\cap\mathbf{Z}(G)} = \chi_j$ for some j , $1 \leq j \leq n_1$. Hence $\lambda = \widetilde{\chi}_j$ on $C_1 \cap \mathbf{Z}(G)$ implies $\operatorname{ind}_{C_1\uparrow G} \lambda \simeq \operatorname{ind}_{C_1\uparrow G} \widetilde{\chi}_j$, i.e. that $\operatorname{ind}_{C_1\uparrow G} \lambda$ is equivalent to some $\operatorname{ind}_{C_1\uparrow G} \widetilde{\chi}_j$, $1 \leq j \leq n_1$. Finally if $C_1 = C$ then $\mathfrak{N}_{C_1} = L_{\mathbf{Z}}$; so the last assertion of Theorem 3.5.5 follows from the first assertion.

Corollary 3.5.6 _Let_ $G = KAN$ _be an Iwasawa decomposition of_ G , _where_ A _is abelian,_ N _is nilpotent and_ K _is compact. Then for any unitary character_ λ _of_ A _we have_

$$\operatorname{ind}_{A\uparrow G} \lambda \simeq \operatorname{ind}_{A\uparrow G} 1$$

where 1 _is the trivial character of_ A .

Proof: In Theorem 3.5.5 take $C_1 = A$. The corollary follows since $A \cap \mathbf{Z}(G) = \{e\}$.

3.6 $\operatorname{ind}_{C\uparrow G} \lambda$ as a subrepresentation of $L^2(A\backslash G)$

Suppose now that the C_1 of C is C. Then we consider the class of induced representations of G

$$U(\cdot,\lambda) = \operatorname{ind}_{C\uparrow G} \lambda$$

as λ varies over $\hat{C}$. In this section we shall characterize $U(\cdot,\lambda)$ as a sub-representation of $L^2(A\backslash G)$. The basic tool needed is the imprimitivity theorem. Theorem 3.5.5 will also play a key role.

Since M centralizes A, $m\,A\,x = A\,M\,x$ for all $(m,x) \in M \times A$. Therefore M acts on $A\backslash G$ on the left. Choose a G-invariant measure $d\dot{x}$ on $A\backslash G$ such that

$$\int_G f(x)dx = \int_{A\backslash G} \int_A f(ax)da\, d\dot{x} \tag{3.6.1}$$

for all $f \in C_c(G)$, where dx, da are right Haar measures on G, A; remember that G, A are unimodular. Since M acts on $A\backslash G$ on the left, it is not clear that $d\dot{x}$ is M-invariant. However $am = ma$ for all $(a,m) \in A \times M$. This is enough to show M-invariance:

Take $k \in C_c(A\backslash G)$, $m \in M$ arbitrary. Choose $\phi \in C_c(G)$ such that

$$k(Ax) = \int_A \phi(ax)da \tag{3.6.2}$$

for all x in G. Then $\int_{A\backslash G} k(m \cdot Ax)d\dot{x} = \int_{A\backslash G} \int_A \phi(amx)da\, d\dot{x} =$

$$= \int_{A\backslash G} \int_A \phi(max)da\, d\dot{x} = \int_G \phi(mx)dx\,, \text{ by (3.6.1)}, = \int_G \phi(x)dx =$$

$$\int_{A\backslash G} k(Ax)d\dot{x}\,, \text{ by (3.6.1) and (3.6.2).}$$

The space of orbits $M\backslash(A\backslash G)$ is easily computed. One uses the Iwasawa decomposition $G = ANK$.

Proposition 3.6.3 The map $Cx \longrightarrow$ orbit of $An(x)\,k(x)$ is a (well-defined) homeomorphism of $C\backslash G$ onto $M\backslash(A\backslash G)$.

By Proposition 3.6.3 we have the following situation: M is a compact group acting on the space $A\backslash G$ which has an M-invariant measure $d\dot{x}$. The space of orbits is $C\backslash G$. This implies the following result.

Proposition 3.6.4 Let dm denote the normalized Haar measure on M. Then there is a unique measure $d\bar{x}$ on $C\backslash G$ such that

$$\int_{A\backslash G} f(Ax)d\dot{x} = \int_{C\backslash G} \int_M f\bigl(Amn(x)\,k(x)\bigr)\, dm\, d\bar{x} \tag{3.6.5}$$

for all $f \in C_c(A\backslash G)$. Moreover $d\bar{x}$ is G-invariant (since $d\dot{x}$ is also G-invariant).

Given $\lambda \in \hat{C}$, let $H(\lambda)$ be the representation space of $U(\circ, \lambda)$: $H(\lambda)$ is the space of measurable functions $f : G \longrightarrow \mathbb{C}$

such that

(i) $$f(cx) = \lambda(c)\, f(x)$$

(ii) $$\int_{C\backslash G} |f(x)|^2\, d\bar{x} < \infty$$

for all $(c, x) \in C \times G$. We set

$$L_\lambda^2(A\backslash G) = \left\{f \in L^2(A\backslash G) \mid f(m \cdot Ax) = \lambda(m)\, f(Ax)\ \forall\ (m, x) \in M \times G\right\}.$$

$L_\lambda^2(A\backslash G)$ is a closed subspace of $L^2(A\backslash G)$.

Proposition 3.6.5 The formula

$$(\Phi_\lambda f)(Ax) = f\Big(n(x)\, k(x)\Big),$$

where $(f, x) \in H(\lambda) \times G$, defines a unitary map Φ_λ of $H(\lambda)$ onto $L_\lambda^2(A\backslash G)$. The transform of $U(\cdot, \lambda)$ under Φ acts on the space $L_\lambda^2(A\backslash G)$ as follows:

$$\Big(U(g, \lambda)\, f\Big)(Ax) = \lambda\Big(a\big(k(x)\, g\big)\Big)\, f(A\, x\, g),$$

where $$(f, x, g) \in L_\lambda^2(A\backslash G) \times G \times G.$$

Proof: Φ_λ is well-defined: Suppose $Ax = Ay$ so $ax = y$, $a \in A$. Then $a\, a(x)\, n(x)\, k(x) = a(y)\, n(y)\, k(y)$; hence $a\, a(x) = a(y)$, $n(x) = n(y)$ and $k(x) = k(y)$ so that $f\big(n(x)\, k(x)\big) = f\big(n(y)\, k(y)\big)$. Clearly $(\Phi f)(m \cdot Ax) = \Phi f(Amx) = f\big(n(mx)\, k(mx)\big)$. But $mx = m\, a(x)\, n(x)\, k(x) = a(x)\, m\, n(x) m^{-1}\, m\, k(x) \Rightarrow a(mx) = a(x)$, $n(mx) =$

$m\, n(x) m^{-1}$, $k(mx) = m\, k(x)$. Therefore $(\Phi_\lambda f)(m \cdot Ax) = f\big(m\, n(x)\, k(x)\big) = \lambda(m)\, \Phi_\lambda f\,(Ax)$, since $(m, f) \in C \times H(\lambda)$. By Proposition 3.6.4

$$\int_{A\backslash G} |\Phi_\lambda f|^2\, d\dot{x} = \int_{C\backslash G} \int_M \left|\Phi_\lambda f\big(Amn(x)\, k(x)\big)\right|^2 dm\, d\bar{x}$$

$$= \int_{C\backslash G} \int_M |\lambda(m)|^2 \left|f\big(n(x)\, k(x)\big)\right|^2 dm\, d\bar{x}$$

$$= \int_{C\backslash G} |f(x)|^2\, d\bar{x} < \infty$$

$\Big($since $f(x) = \lambda\big(a(x)\big)\, f\big(n(x)\, k(x)\big)\Big)$. Therefore $\Phi_\lambda f \in L^2_\lambda(A\backslash G)$. If $h \in L^2_\lambda(A\backslash G)$, $\Phi^{-1} h$ is given by $(\Phi_\lambda^{-1} h)(x) = \lambda\big(a(x)\big)\, h(Ax)$. If $(g, x, h) \in G \times G \times L^2_\lambda(A\backslash G)$, then

$$n(x)\, k(x) g = a\big(k(x)\, g\big)\, a\big(k(x)\, g\big)^{-1}\, n(x)\, a\big(k(x)\, g\big)\, n\big(k(x)\, g\big)\, k\big(k(x)\, g\big).$$

Therefore $a\big(n(x)\, k(x)\, g\big) = a\big(k(x)\, g\big)$ so that

$$\big(\Phi_\lambda\, U(g, \lambda)\, \Phi_\lambda^{-1} h\big)(Ax) = \big(U(g, \lambda)\, \Phi_\lambda^{-1} h\big)\big(n(x)\, k(x)\big)$$

$$= \big(\Phi_\lambda^{-1} h\big)\big(n(x)\, k(x) g\big)$$

$$= \lambda\Big(a\big(k(x) g\big)\Big)\, h\big(A\, n(x)\, k(x) g\big)$$

$$= \lambda\Big(a\big(k(x) g\big)\Big)\, h(A\, x\, g).$$

Now let $R^{A\backslash G}$ denote the right regular representation of G on $L^2(A\backslash G)$ defined by the G-invariant measure $d\dot{x}$. The subspace

$L^2_\lambda(A\backslash G)$ is $R^{A\backslash G}$ invariant. If Y is any topological space, let $\mathcal{B}(Y)$ denote the σ-algebra of Borel subsets of Y. We define a spectral measure P on $A\backslash G$ with values in the set of projections on $L^2(A\backslash G)$ by setting

$$P(\sigma)f = \chi_\sigma f$$

for $\sigma \in \mathcal{B}(A\backslash G)$, $f \in L^2(A\backslash G)$, where χ_σ is the characteristic function of σ. Think of $R^{A\backslash G}$ as $\operatorname{ind}_{A\uparrow G} 1$. Then by 2.4, P is the spectral measure canonically associated with the induced representation $R^{A\backslash G}$.

Lemma 3.6.6 If $\sigma \in \mathcal{B}(A\backslash G)$ is such that $m \cdot \sigma = \sigma$ for all $m \in M$, then $L^2_\lambda(A\backslash G)$ is $P(\sigma)$ invariant.

Define $j : A\backslash G \longrightarrow C\backslash G$ by $j(Ax) = Cx$, $x \in C$. j is well-defined, continuous and onto. If π_C, π_A are the projections of G onto $C\backslash G$, $A\backslash G$, respectively, then we have the commutative diagram:

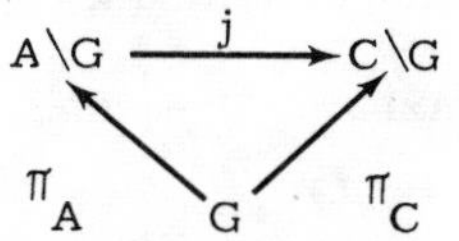

If $\tau \in \mathcal{B}(C\backslash G)$ then clearly $j^{-1}(\tau) \in \mathcal{B}(A\backslash G)$.

Lemma 3.6.7 For all $\tau \in \mathcal{B}(A\backslash G)$, $j^{-1}(\tau)$ satisfies Lemma 3.6.6 ; hence $L^2_\lambda(A\backslash G)$ is $P\left(j^{-1}(\tau)\right)$ invariant.

Note that $P \circ j^{-1}$ is a spectral measure on $C\backslash G$ with values in the set of projections on $L^2(A\backslash G)$. Let $P(\cdot, \lambda)$ be the canonical spectral measure on $C\backslash G$, with values in the set of projections on $H(\lambda)$ associated with the induced representation $U(\cdot, \lambda)$; see 2.4 again.

THEOREM 3.6.8 $\Phi_\lambda P(\tau, \lambda) \Phi_\lambda^{-1} = P(j^{-1}(\tau))\Big|_{L^2_\lambda(A\backslash G)}$ for all $\tau \in \mathcal{B}(C\backslash G)$, where Φ_λ is defined in Proposition 3.6.5.

Proof: By Lemma 3.6.7 $L^2_\lambda(A\backslash G)$ is $P\left(j^{-1}(\tau)\right)$ invariant. Take $(h, x) \in L^2_\lambda(A\backslash G) \times G$. If $h = \Phi f$, $f \in H(\lambda)$, then

$$\begin{aligned}\left(\Phi_\lambda P(\tau, \lambda) \Phi_\lambda^{-1} h\right)(Ax) &= \left(P(\tau, \lambda)\, f\right)\left(n(x)\, k(x)\right)\\ &= \chi_\tau\left(C\, n(x)\, k(x)\right) f\left(n(x)\, k(x)\right)\\ &= \chi_\tau\left(C\, n(x)\, k(x)\right) h(Ax)\ .\end{aligned}$$

On the other hand, $\left(P\left(j^{-1}(\tau)\right) h\right)(Ax) = \chi_{j^{-1}(\tau)}(Ax)\, h(Ax)$. But $C\, n(x)\, k(x) = j(Ax) \in \tau$ if and only if $Ax \in j^{-1}(\tau)$; so $\chi_\tau\left(C\, n(x)\, k(x)\right) = \chi_{j^{-1}(\tau)}(Ax)$.

If we set $\lambda_o(ma) = \lambda(m)$, $ma \in C = MA$, then $\lambda_o \in \hat{C}$ and λ_o is trivial on A. Since $\lambda_o = \lambda$ on M, $\lambda_o = \lambda$ on $\mathcal{Z}(G) \subset M$ so $U(\cdot, \lambda) \simeq U(\cdot, \lambda_o)$ by Theorem 3.5.5. By Proposition 3.6.5 $U(\cdot, \lambda_o)$

acts on $L_{\lambda_o}^2(A\backslash G) \equiv L_\lambda^2(A\backslash G)$ by right translation. Thus

$$U(\cdot, \lambda_o) = R^{A\backslash G}\Big|_{L_\lambda^2(A\backslash G)}$$

so we have that $U(\cdot, \lambda)$ is a sub-representation of $R^{A\backslash G}$ on $L_\lambda^2(A\backslash G)$. By Proposition 3.6.5 and Theorem 3.6.8

$$\left(U(\cdot, \lambda_o), P(\cdot, \lambda_o)\right) \overset{\Phi_{\lambda_o}}{\simeq} \left(R^{A\backslash G}\Big|_{L_{\lambda_o}^2(A\backslash G)}, \ P \circ j^{-1}\Big|_{L_{\lambda_o}^2(A\backslash G)}\right).$$

Therefore the pair $\left(R^{A\backslash G}, P \circ j^{-1}\right)\Big|_{L_\lambda^2(A\backslash G)}$ is irreducible because $\left(U(\cdot, \lambda_o), P(\cdot, \lambda_o)\right)$ is irreducible by Theorem 2.4.2 (λ_o is irreducible).

Conversely, suppose $\mathscr{K} \neq 0$ is an arbitrary $\left(R^{A\backslash G}, P \circ j^{-1}\right)$ invariant, $\left(R^{A\backslash G}, P \circ j^{-1}\right)$ irreducible subspace of $L^2(A\backslash G)$. As $\mathscr{K}$ is $P \circ j^{-1}$ invariant, $P \circ j^{-1}\Big|_{\mathscr{K}}$ is a spectral measure on $C\backslash G$ with values in the set of projections on $\mathscr{K}$.

Proposition 3.6.9 Define $P_{\mathscr{K}} = P \circ j^{-1}\Big|_{\mathscr{K}}$. Then $P_{\mathscr{K}}$ is a system of imprimitivity for $R^{A\backslash G}\Big|_{\mathscr{K}}$ based on the action of G on $C\backslash G$; i.e., for all $(x, \tau, \theta) \in G \times \mathscr{B}(C\backslash G) \times \mathscr{K}$

$$R^{A\backslash G}(x)\, P_{\mathscr{K}}(\tau)\, R^{A\backslash G}(x^{-1})\theta = P_{\mathscr{K}}(\tau \cdot x^{-1})\theta \ ;$$

see Definition 2.4.3.

Proof: For all $y \in G$,

$$\left(R^{A\backslash G}(x)\, P_{\mathscr{X}}(\tau)\, R^{A\backslash G}(x^{-1})\theta\right)(Ay) = \left(P_{\mathscr{X}}(\tau)\, R^{A\backslash G}(x^{-1})\theta\right)(Ayx)$$

$$= \chi_{j^{-1}(\tau)}(Ayx)\left(R^{A\backslash G}(x^{-1})\theta\right)(Ayx)$$

$$= \chi_{j^{-1}(\tau)}(Ayx)\, \theta(Ay)\,.$$

On the other hand,

$$\left(P_{\mathscr{X}}(\tau \cdot x^{-1})\theta\right)(Ay) = \chi_{j^{-1}(\tau \cdot x^{-1})}(Ay)\, \theta(Ay)\,.$$

But $Ay \in j^{-1}(\tau \cdot x^{-1})$ if and only if $Cy = j(Ay) \in \tau \cdot x^{-1}$ which holds if and only if $Cyx \in \tau$ and this is true if and only if $Ayx \in j^{-1}(\tau)$; hence,

$$\chi_{j^{-1}(\tau)}(Ayx) = \chi_{j^{-1}(\tau \cdot x^{-1})}(Ay)\,.$$

According to Theorem 2.4.4 (the imprimitivity theorem), there is a unique (up to equivalence) unitary representation λ of C such that

$$\left(\operatorname*{ind}_{C\uparrow G} \lambda\,,\ P(\cdot, \lambda)\right) \simeq \left(R^{A\backslash G}\big|_{\mathscr{X}}\,,\ P_{\mathscr{X}}\right).$$

By hypothesis $\left(R^{A\backslash G}\big|_{\mathscr{X}}\,, P_{\mathscr{X}}\right)$ is irreducible. Therefore, $\left(\operatorname*{ind}_{C\uparrow G} \lambda,\ P(\cdot, \lambda)\right)$ is irreducible. Applying Theorem 2.4.2, we find that λ is irreducible. But an irreducible representation of an abelian group is 1-dimensional! Hence $\lambda \in \hat{C}$.

In summary we have obtained the following

THEOREM 3.6.10 *Let* $\lambda \in \hat{C}$. *The subspace* $L^2_\lambda(A\backslash G)$ *of* $L^2(A\backslash G)$ *is* $(R^{A\backslash G}, P \circ j^{-1})$ *invariant and* $(R^{A\backslash G}, P \circ j^{-1})$ *irreducible.* $(R^{A\backslash G}, P \circ j^{-1})\Big|_{L^2_\lambda(A\backslash G)} \simeq (U(\cdot, \lambda_o), P(\cdot, \lambda_o))$ *for a unique* $\lambda_o \in \hat{C}$. *Moreover* $U(\cdot, \lambda_o) \simeq U(\cdot, \lambda)$ *so that* $U(\cdot, \lambda)$ *is a sub-representation of* $R^{A\backslash G}$. *Conversely, suppose* T *is a sub-representation of* $R^{A\backslash G}$ *which acts on a* $(R^{A\backslash G}, P \circ j^{-1})$ *invariant,* $(R^{A\backslash G}, P \circ j^{-1})$ *irreducible subspace* $\mathcal{H}_T$ *of* $L^2(A\backslash G)$. *Then* $P \circ j^{-1}\Big|_{\mathcal{H}_T}$ *is a system of imprimitivity for* T *based on the action of* G *on* $C\backslash G$ *and there exists a unique* $\lambda \in \hat{C}$ *such that* $\left(T, P \circ j^{-1}\Big|_{\mathcal{H}_T}\right) \simeq (U(\cdot, \lambda), P(\cdot, \lambda))$. *In particular* $T \simeq U(\cdot, \lambda)$.

CHAPTER 4

The Tensor Product of Principal Series Representations of a Complex Semi-Simple Lie Group

Recently N. Wallach, [33], using strong results of B. Kostant, [18], has shown, once and for all, that every non-degenerate principal series representation of a complex semi-simple Lie group is irreducible. It is rather natural then to analyze the tensor product of two such representations.

For the group $G = SL(2,\mathbb{C})$, results due to G. Mackey, [26], and M. Naimark, [30], are available. Naimark's paper is very explicit. N. Anh, [1], considers the group $G = SL(n,\mathbb{C})$. His techniques, like those of Mackey, exploit the irreducibility of the restriction of the principal series of the subgroup of matrices whose top row vector is $(1,0,\ldots,0)$. Such strong irreducibility phenomenon is limited to the unimodular group. It is clear then that in considering the general complex Lie group one must find other methods of analyzing the tensor product.

As indicated in the introduction, we shall decomposed the restriction of the principal series to the Cartan subgroup and then apply the Frobenius-Mackey Reciprocity Theorem.

4.1 Construction of intertwining operators

We retain the notation established in Chapter 3. Let $\hat{B}$ be the set of 1-dimensional unitary representations of B.

Definition 4.1.1 *The non-degenerate principal series of unitary representations of the connected complex semi-simple Lie group* G *is the family of induced representations* $\operatorname{ind}_{B\uparrow G} \lambda$, *where* λ *varies over* $\hat{B}$.

We shall write $S(\cdot,\lambda) = \operatorname{ind}_{B\uparrow G} \lambda$, $\lambda \in \hat{B}$, and refer to $S(\cdot,\lambda)$, more simply, as a principal series representation of G.

THEOREM 4.1.2 *Every* $S(\cdot,\lambda)$ *is irreducible.*

A proof is given in [32].

$S(\cdot,\lambda)$ is easily realized as a multiplier representation on $L^2(V)$, where V is the maximal nilpotent subgroup of G of Section

3.1. This essentially follows from the fact, indicated earlier, that $G = BV$, up to a set of measure zero. One also uses the integration formula (3.1.24). The result is

THEOREM 4.1.3 The principal series element $S(\cdot, \lambda)$ acts on the space $L^2(V)$ by:

$$\left(S(a, \lambda) f\right)(v) = \lambda\left(b(va)\right) \mu^{\frac{1}{2}}\left(b(va)\right) f\left(v(va)\right) ,$$

where $(a, v) \in G \times V$, $va = b(va)\, v(va) \in BV$, and $\mu = \frac{1}{\Delta_B}$.

Refer to [19] for a proof.

From Theorem 4.1.3 it follows that if $\lambda_1, \lambda_2 \in \hat{B}$, then the tensor product $S(\cdot, \lambda_1) \otimes S(\cdot, \lambda_2)$ of $S(\cdot, \lambda_1)$ and $S(\cdot, \lambda_2)$ can be realized as the unitary representation of G on $L^2(V \times V)$ given by the formula:

(4.1.4) $$\left(S(a, \lambda_1) \otimes S(a, \lambda_2) f\right)(v_1, v_2)$$

$$= \lambda_1\left(b(v_1 a)\right) \mu^{\frac{1}{2}}\left(b(v_1 a)\right) \lambda_2\left(b(v_2 a)\right) \mu^{\frac{1}{2}}\left(b(v_2 a)\right) f\left(v(v_1 a), v(v_2 a)\right) ,$$

$(a, v_1, v_2) \in G \times V \times V$, $f \in L^2(V \times V)$, $\mu = \frac{1}{\Delta_B}$

Since $S(\cdot, \lambda_1) \otimes S(\cdot, \lambda_2)$ is a tensor product of induced representations we can begin the study of $S(\cdot,\lambda_1)\otimes S(\cdot,\lambda_2)$ by applying Mackey's Theorem 2.2.8. Here we take $H_1 = H_2 = B$. By Bruhat's lemma (Theorem 3.1.11), the number of $B:B$ double cosets is actually finite and only one of them Bp_oB, $p_o \in M'$, has positive Haar measure. Therefore the sum $\sum_d^{\oplus} V^d$ in Theorem 2.2.8 reduces to a single summand $V^d = V^{e,p_o}$. $\lambda_1^e(b_1) = \lambda_1(b_1)$, $\lambda^{p_o^{-1}}(b_2) = \lambda_2\left(p_o^{-1} b_2 p_o\right)$ for all $(b_1, b_2) \in B \times p_o B p_o^{-1}$. In our case

$$H_{e,p_o^{-1}} = e^{-1} Be \cap p_o B p_o^{-1}$$

$$= B \cap p_o C p_o^{-1} p_o N p_o^{-1} = B \cap CV$$

(see (3.1.16)). (Note that $p_o C p_o^{-1} = p_o M p_o^{-1} p_o A p_o^{-1} = MA = C$ since $p_o \in M'$, M is normal in M' and M' is the normalizer of A in K.) Now $B \cap CV = C$ since elements in CNV are written uniquely. Thus $H_{e,p_o^{-1}} = C$. Finally, N is the commutator subgroup of B; so any character λ of B is the extension of a character λ of C such that $\lambda(N) = 1$. We have $T^{e,p_o^{-1}} = \lambda_1\big|_C \otimes \lambda_2^{p_o^{-1}}\big|_C$. Therefore, by Theorem 2.2.8

$$S(\cdot\,,\lambda_1)\otimes S(\cdot\,,\lambda_2)\simeq \operatorname*{ind}_{C\uparrow G} \lambda_1\Big|_C \otimes \lambda_2^{p_o^{-1}}\Big|_C \ .$$

This proves

Proposition 4.1.5 <u>Let</u> $\lambda_1, \lambda_2 \in \hat{C}$. <u>Extend</u> λ_1, λ_2 <u>to</u> B <u>by requiring triviality on</u> N . <u>Let</u> $p_o \in M'$ <u>be the "Bruhat element" of Theorem</u> 3.1.11. <u>Then</u>

$$S(\cdot\,,\lambda_1)\otimes S(\cdot\,,\lambda_2)\simeq \operatorname*{ind}_{C\uparrow G} \lambda_1(p_o\lambda_2) \ ,$$

<u>where</u> $p_o\lambda(c) = \lambda(p_o^{-1} c\, p_o)$ <u>for any</u> $(c,\lambda)\in C\times\hat{C}$.

From Theorem 3.3.5, taking $C_1 = C$, we deduce

Proposition 4.1.6 $S(\cdot\,,\lambda_1)\otimes S(\cdot\,,\lambda_2)$ <u>can be realized as a representation of</u> G <u>on</u> $L^2(N\times V)$ <u>as follows:</u>

$$\left(S(a,\lambda_1)\otimes S(a,\lambda_2) f\right)(n,v) =$$

(4.1.6) $$\lambda_1\left(c(va)\right)\lambda_2\left(p_o^{-1}c(va)p_o\right) f\left(c(va)^{-1} n\, c(va)\, n(va),\ v(va)\right) \ ,$$

$(f,n,v,a)\in L^2(N\times V)\times N\times V\times G$.

Of course the Mackey theory does not give an explicit unitary operator which defines the equivalence in Proposition 4.1.5. We will construct such an operator for general complex semi-simple Lie groups. The construction is based on certain "matrix identities." We will need to observe:

THEOREM 4.1.7 $p_o^2 \in M$.

Proof: If $\Pi = \{\alpha_1, \cdots, \alpha_\ell\}$ is the system of simple roots, then by (3.1.15) there is a permutation $(n(1), \cdots, n(\ell))$ of the letters $1, 2, \cdots, \ell$ such that $p_o \alpha_j = -\alpha_{n(j)}$, $1 \le j \le \ell$. Therefore $p_o^2 \Pi = \Pi$. But the Weyl group $W = M'/M$ acts simply transitively on the simple root systems; see [17], page 242. Hence $p_o^2 = 1$ in W .

Corollary 4.1.8 $V = p_o^{-1} N p_o$ (cf. (3.1.16)) .

Proof: M normalizes V and N .

Lemma 4.1.9 Let $(a, v, n) \in G \times V \times N$. Then the following equations, when both sides are well-defined, are valid:

(i) $b\left[p_o^{-1} c(va)^{-1} n\, c(va)\, n(va)\right] = \left(p_o^{-1} c(va)^{-1} p_o\right) b(p_o^{-1} n)\, b\left[v(p_o^{-1} n)\, va\right]$

(ii) $v\left[p_o^{-1} c(va)^{-1} n\, c(va)\, n(va)\right] = v\left[v(p_o^{-1} n)\, va\right] v(va)^{-1}$

(iii) $c\left[p_o^{-1} c(va)^{-1} n\, c(va)\, n(va)\right] = \left(p_o^{-1} c(va)^{-1} p_o\right) c(p_o^{-1} n)\, c\left[v(p_o^{-1} n)\, va\right]$

(iv) $n\left[p_o^{-1} c(va)^{-1} n\, c(va)\, n(va)\right]$

$$= c\left[v(p_o^{-1} n)\, va\right]^{-1} n(p_o^{-1} n)\, c\left[v(p_o^{-1} n)\, va\right] n\left[v(p_o^{-1} n)\, va\right]$$

(v) $b(p_o^{-1} n)\, b\left[v(p_o^{-1} n)\, p^{-1}\right] = p_o^{-2}, \quad v\left[v(p_o^{-1} n)\, p_o^{-1}\right] = p_o\, n\, p_o^{-1}$

(vi) $b(v\, p_o^{-1})\, b\left[v(v\, p_o^{-1})\, p_o\right] = e$

(vii) $v\left[v(v\, p^{-1})\, p_o\right] = v$

(viii) $p_o^{-1}\, v\left[v(p_o^{-1} n)\, p_o^{-1}\right] p_o = n$.

<u>Proof</u>: For notational simplicity we shall write p instead of p_o throughout the proof. We have

$p^{-1} c(va)^{-1} n\, c(va)\, n(va)$

$= p^{-1} c(va)^{-1} p\, p^{-1} n\, c(va)\, n(va)\, c(va)^{-1} c(va)$

$= \left(p^{-1} c(va)^{-1} p\right) b\left[p^{-1} n c(va) n(va) c(va)^{-1}\right] c(va) c(va)^{-1} v\left[p^{-1} n c(va) n(va) c(va^{-1}\right] c(va)$

$\Rightarrow$ a. $b\left[p^{-1} c(va)^{-1} n\, c(va)\, n(va)\right]$

$= \left(p^{-1} c(va)^{-1} p\right) b\left[p^{-1} n\, c(va)\, n(va)\, c(va)^{-1}\right] c(va)$ and

b. $v\left[p^{-1} c(va)^{-1} n\, c(va)\, n(va)\right]$

$= c(va)^{-1}\, v\left[p^{-1} n\, c(va)\, n(va)\, c(va)^{-1}\right] c(va)$.

On the other hand

$$va = c(va)\,n(va)\,v(va) \text{ implies } c(va)\,n(va)$$

$$= va\,v(va)^{-1} \text{ which implies } p^{-1}n\,c(va)\,n(va)\,c(va)^{-1}$$

$$= b(p^{-1}n)\,v(p^{-1}n)\,va\,v(va)^{-1}\,c(va)^{-1}$$

$$= b(p^{-1}n)\,b\left[v(p^{-1}n)va\right]c(va)^{-1}\,c(va)\,v\left[v(p^{-1}n)va\right]v(va)^{-1}\,c(va)^{-1} \quad ,$$

so that

$$\text{c.} \quad b\left[p^{-1}n\,c(va)\,n(va)\,c(va)^{-1}\right] = b(p^{-1}n)\,b\left[v(p^{-1}n)\,va\right]c(va)^{-1} \quad ,$$

$$\text{d.} \quad v\left[p^{-1}n\,c(va)\,n(va)\,c(va)^{-1}\right] = c(va)\,v\left[v(p^{-1}n)\,va\right]v(va)^{-1}\,c(va)^{-1} \; .$$

a. and c. imply (i) while b. and d. imply (ii) . From a. ,

$$c\left[p^{-1}\,c(va)^{-1}n\,c(va)\,n(va)\right]\,n\left[p^{-1}\,c(va)^{-1}n\,c(va)\,n(va)\right]$$

$$= \left(p^{-1}c(va)^{-1}p\right)c\left[p^{-1}nc(va)n(va)c(va)^{-1}\right]c(va)c(va)^{-1}n\left[p^{-1}nc(va)n(va)c(va)^{-1}\right]c(va)$$

so that

$$\text{e.} \quad c\left[p^{-1}\,c(va)^{-1}n\,c(va)\,n(va)\right]$$

$$= \left(p^{-1}\,c(va)^{-1}p\right)c\left[p^{-1}n\,c(va)\,n(va)\,c(va)^{-1}\right]c(va)$$

$$\text{f.} \quad n\left[p^{-1}\,c(va)^{-1}n\,c(va)\,n(va)\right] = c(va)^{-1}n\left[p^{-1}n(c(va)n(va)c(va)^{-1}\right]c(va) \, .$$

From c .

$$c\left[p^{-1} n\, c(va)\, n(va)\, c(va)^{-1}\right] n\left[p^{-1} n\, c(va)\, n(va)\, c(va)^{-1}\right]$$

$$= c(p^{-1}n)\, n(p^{-1}n)\; c\left[v(p^{-1}n)\, va\right] n\left[v(p^{-1}n)\, va\right]\; c(va)^{-1}$$

$$= \left[c(p^{-1}n)\, c[\, v(p^{-1}n)\, va]\quad (va)^{-1}\right]$$

$$\left\{c(va)\left(c[v(p^{-1}n)\, va]^{-1}\, n(p^{-1}n)\, c[v(p^{-1}n)\, va]\right)\, c(va)^{-1}\right\}$$

$$\left[c(va)\; n[v(p^{-1}n)\, va]\, c(va)^{-1}\right] \quad ;$$

hence

g. $$c\left[p^{-1} n\, c(va)\, n(va)\, c(va)^{-1}\right] = c(p^{-1}n)\, c\left[v(p^{-1}n)\, va\right]\; c(va)^{-1} \quad ,$$

h. $$n\left[p^{-1} n\, c(va)\, n(va)\, c(va)^{-1}\right]$$

$$= c(va)\, c\left[v(p^{-1}n)\, va\right]^{-1} n(p^{-1}n)\; c\left[v(p^{-1}n) va\right] n\left[v(p^{-1}n)\, va\right]\; c(va)^{-1} .$$

e. and g. imply (iii), while f. and h. imply (iv). Next let $p^2 = m \in M$, (by Theorem 4.1.7). Now

$$\begin{aligned} p^{-1}n &= b(p^{-1}n)\, v(p^{-1}n)\, p^{-1}\, p \\ &= b(p^{-1}n)\, b\left[v(p^{-1}n)\, p^{-1}\right] v\left[v(p^{-1}n)\, p^{-1}\right] p \ . \end{aligned}$$

Thus

$$p^{-1}n\, p = b(p^{-1}n)\; b\left[v(p^{-1}n)\; p^{-1}\right] m\, m^{-1}\, v\left[\, v(p^{-1}n)\; p^{-1}\right]\, m \quad .$$

But $p^{-1} n\, p \in V$ by Corollary 4.1.8 so that $b(p^{-1}n)\, b\left[v(p^{-1}n)\, p^{-1}\right] m = e$

and $m^{-1} v\left[v(p^{-1}n)p^{-1}\right] m = p^{-1} n p$; this implies (v). To prove (vi) and (vii) note that

$$vp^{-1} = b(vp^{-1})\, v(vp^{-1}) p\, p^{-1} = b(vp^{-1})\, b\left[v(vp^{-1})\, p\right] v\left[v(vp^{-1})\, p\right] p^{-1}$$

and hence that

$$v = b(vp^{-1})\, b\left[v(vp^{-1})\, p\right] v\left[v(vp^{-1})\, p\right] .$$

Finally observe that (v) implies (viii).

THEOREM 4.1.10 Let G be a connected complex semi-simple Lie group. Let λ_1, λ_2 be unitary characters of the Cartan subgroup $C = MA$. Lift λ_1, λ_2 to $B = MAN$ by requiring triviality on $N = [B, B]$. If $S(\cdot, \lambda_j)$, $j = 1, 2$, are the corresponding (non-degenerate) principal series elements, then $S(\cdot, \lambda_1) \otimes S(\cdot, \lambda_2) \simeq \operatorname{ind}_{C \uparrow G} \lambda_1 \otimes p_o \lambda_2$, where p_o is the Bruhat element of Theorem 3.1.11. Suppose that $S(\cdot, \lambda_1) \otimes S(\cdot, \lambda_2)$ acts on $L^2(V \times V)$ according to formula (4.1.4) and suppose that $\operatorname{ind}_{C \uparrow G} \lambda_1 \otimes p_o \lambda_2$ acts on $L^2(N \times V)$ according to (4.1.6). Then a unitary operator $\Phi_{\lambda_1, \lambda_2} : L^2(V \times V)$ onto $L^2(N \times V)$ which defines the equivalence of $S(\cdot, \lambda_1) \otimes S(\cdot, \lambda_2)$ and

$\operatorname{ind}_{C\uparrow G} \lambda_1 \otimes p_o \lambda_2$ is given as follows:

$$\left(\Phi_{\lambda_1, \lambda_2} f\right)(n, v) = \mu^{\frac{1}{2}}\left(b(p_o^{-1} n)\right) \lambda_2\left(b(p_o^{-1} n)\right) f\left(v, v(p_o^{-1} n) v\right) \quad ,$$

$(f, n, v) \in L^2(V \times V) \times N \times V$, $\mu = \frac{1}{\Delta_B}$ (note that $\Phi_{\lambda_1, \lambda_2}$ is independent of λ_1). $\Phi^{-1}_{\lambda_1, \lambda_2} : L^2(N \times V) \longrightarrow L^2(V \times V)$ is given by

$$\left(\Phi^{-1}_{\lambda_1, \lambda_2} g\right)(v_1, v_2) = \mu^{\frac{1}{2}}\left(b(v_2 v_1^{-1} p_o^{-1})\right) \lambda_2\left(b(v_2 v_1^{-1} p_o^{-1})\right) \lambda_2(p_o^2) \cdot$$

$$g\left(p_o^{-1} v\left[v_2 v_1^{-1} p_o^{-1}\right] p_o, v_1\right) \quad , \quad (g, v_1, v_2) \in L^2(N \times V) \times V \times V \ .$$

Proof: Let us assume for the moment that $\Phi_{\lambda_1, \lambda_2}$ is indeed unitary. For $f \in L^2(V \times V)$, $(a, n, v) \in G \times N \times V$ we have

$$\left(\Phi_{\lambda_1, \lambda_2} S(a, \lambda_1) \otimes S(a, \lambda_2) f\right)(n, v)$$
$$= \mu^{\frac{1}{2}}\left(b(p_o^{-1} n)\right) \lambda_2\left(b(p_o^{-1} n)\right) \left(S(a, \lambda_1) \otimes S(a, \lambda_2) f\right) \left(v, v(p_o^{-1} n) v\right) \ .$$

Now

$$\left(S(a, \lambda_1) \otimes S(a, \lambda_2) f\right)\left(v, v(p_o^{-1} n) v\right)$$
$$= \mu^{\frac{1}{2}}\left(b(va)\right) \lambda_1\left(b(va)\right) \mu^{\frac{1}{2}}\left(b\left[v(p_o^{-1} n) va\right]\right) \lambda_2\left(b\left[v(p_o^{-1} n) va\right]\right) \cdot$$

$f\left(v(va),\ v\left[v(p_o^{-1}n)\,va\right]\right)$, by (4.1.4). On the other hand, by (4.1.6)

$$\underset{C\uparrow G}{\mathrm{ind}}\left(\lambda_1 \otimes p_o\lambda_2\,(a)\,\Phi_{\lambda_1,\lambda_2} f\right)(n, v)$$

$$= \lambda_1\big(c(va)\big)\,\lambda_2\big(p_o^{-1}\,c(va)p_o\big)\ \Phi_{\lambda_1,\lambda_2} f\ \left(c(va)^{-1} n\, c(va)\, n(va),\ v(va)\right)$$

while

$$\Phi_{\lambda_1,\lambda_2} f\ \left(c(va)^{-1} n\, c(va)\, n(va),\ v(va)\right)$$

$$= \mu^{\frac{1}{2}}\left(b\left[p_o^{-1}\,c(va)^{-1} n\, c(va)\, n(va)\right]\right)\lambda_2\left(b\left[p_o^{-1}\,c(va)^{-1} n\, c(va)\, n(va)\right]\right)$$

$$f\left(v(va),\ v\left[\,p_o^{-1}\,c(va)^{-1} n\, c(va)\, n(va)\right] v(va)\right)\ .$$

By Lemma 4.1.7,

$$v\left[p_o^{-1}\,c(va)^{-1} n\, c(va)\, n(va)\right] v(va) = v\left[v(p_o^{-1}n)\ va\right] \qquad \text{((ii))},$$

$$b(p_o^{-1}n)\, b\left[v(p_o^{-1}n)\, va\right] = \left(p_o^{-1}\,c(va)\,p_o\right)\, b\left[p_o^{-1}\,c(va)^{-1} n\, c(va)\, n(va)\right] \qquad \text{((i))},$$

and since $\mu\left(p_o^{-1}\,c\,p_o\right) = \mu(c^{-1})$ for all $c \in C$, by (3.1.26), we get

$$\Phi_{\lambda_1,\lambda_2} S(a,\lambda_1) \otimes S(a,\lambda_2) = \underset{C\uparrow G}{\mathrm{ind}}\ \lambda_1 \otimes p_o\lambda_2\,(a)\,\Phi_{\lambda_1,\lambda_2}$$

for all $a \in G$. Moreover for all $f \in L^2(V \times V)$

$$\left(\Phi^{-1}_{\lambda_1,\lambda_2}\Phi_{\lambda_1,\lambda_2} f\right)(v_1, v_2)$$

$$= \mu^{\frac{1}{2}}\left(b\left(v_2v_1^{-1}p_o^{-1}\right)\right)\lambda_2\left(b\left(v_2v_1^{-1}p_o^{-1}\right)\right)\lambda_2\left(p_o^2\right)\left(\Phi_{\lambda_1,\lambda_2} f\right)$$

$$\left(p^{-1}v\left[v_2v_1^{-1}p_o^{-1}\right]p_o, v_1\right)$$

and

$$\left(\Phi_{\lambda_1,\lambda_2} f\right)\left(p_o^{-1}v\left[v_2v_1^{-1}p_o^{-1}\right]p_o, v_1\right)$$

$$= \mu^{\frac{1}{2}}\left(b\left\{p_o^{-1}p_o^{-1}v\left[v_2v_1^{-1}p_o^{-1}\right]p_o\right\}\right)$$

$$\lambda_2\left(b\left\{p_o^{-1}p_o^{-1}v\left[v_2v_1^{-1}p_o^{-1}\right]p_o\right\}\right)$$

$$f\left(v_1, v\left\{p_o^{-1}p_o^{-1}v\left[v_2v_1^{-1}p_o^{-1}\right]p_o\right\}v_1\right) .$$

Now

$$b\left\{p_o^{-2}v\left[v_2v_1^{-1}p_o^{-1}\right]p_o\right\} = p_o^{-2}b\left\{v\left[v_2v_1^{-1}p_o^{-1}\right]p_o\right\}$$

$$= p_o^{-2}b\left(v_2v_1^{-1}p_o^{-1}\right)^{-1} ,$$

by Lemma 4.1.7 (vi) and the fact that $p_o^{-2} \in M \subset B$. Also

$$v\left\{p_o^{-2}v\left[v_2v_1^{-1}p_o^{-1}\right]p_o\right\}$$

$$= v\left\{v\left[v_2v_1^{-1}p_o^{-1}\right]p_o\right\}$$

$$= v_2v_1^{-1}$$

by Lemma 4.1.7 (vii) and since μ is trivial on M we get

$\Phi^{-1}_{\lambda_1,\lambda_2} \Phi_{\lambda_1,\lambda_2} f = f$. Similarly for all $g \in L^2(N \times V)$

$$\left(\Phi_{\lambda_1,\lambda_2} \Phi^{-1}_{\lambda_1,\lambda_2} g\right)(n, v)$$

$$= \mu^{\frac{1}{2}}\left(b(p_o^{-1}n)\right) \lambda_2\left(b(p_o^{-1}n)\right)\left(\Phi^{-1}_{\lambda_1,\lambda_2} g\right)\left(v, v(p_o^{-1} n) v\right) ,$$

while

$$\left(\Phi^{-1}_{\lambda_1,\lambda_2} g\right)\left(v, v(p_o^{-1} n) v\right)$$

$$= \lambda_2\left(p_o^2\right) \mu^{\frac{1}{2}}\left\{b\left[v\left(p_o^{-1}n\right)v v^{-1} p_o^{-1}\right]\right)\lambda_2\left\{b\left[v\left(p_o^{-1}n\right) v v^{-1} p_o^{-1}\right]\right\} \cdot$$

$$g\left(p_o^{-1} v\left\{\left[v\left(p_o^{-1} n\right) v v^{-1} p_o^{-1}\right]\right\} p_o , v\right) .$$

Now $b\left[v\left(p_o^{-1}n\right) p_o^{-1}\right] = b\left(p_o^{-1}n\right)^{-1} p_o^{-2}$ by Lemma 4.1.7 (v) and

$p_o^{-1} v\left\{\left[v\left(p_o^{-1}n\right) p_o^{-1}\right] p_o = n\right.$ by Lemma 4.1.7 (viii). Since μ is trivial

on M we get $\Phi_{\lambda_1,\lambda_2} \Phi^{-1}_{\lambda_1,\lambda_2} g = g$ so $\Phi^{-1}_{\lambda_1,\lambda_2}$ is indeed the inverse

of $\Phi_{\lambda_1,\lambda_2}$. To show that $\Phi_{\lambda_1,\lambda_2}$ is unitary, we use

Lemma 4.1.11 Given $a \in G$, define $V^a = \{v \in V \mid va \in BV\}$.

Then V^a is an open subset of V , $V - V^a$ has measure zero, and

for any integrable function f on V we have

$$\int_V f(v)\,dv = \int_{V^a} f\Big(v(va)\Big)\,\mu\Big(b(va)\Big)\,dv\ .$$

This is Lemma 28 of [19] .

Since $V = p_o^{-1} N p_o$ we can relate the right Haar measures dv , dn on V, N by the formula

(4.1.10) $$\int_V f(v)\,dv = \int_N f\left(p_o^{-1} n\, p_o\right) dn$$

or

(4.1.11) $$\int_N g(n)\,dn = \int_V g\left(p_o v\, p_o^{-1}\right) dv$$

where f, g are functions on V, N , respectively. Then for $f \in L^2(V\times V)$

$$\int_N\int_V \left|\left(\Phi_{\lambda_1,\lambda_2} f\right)(n,v)\right|^2 dv\,dn = \int_N\int_V \mu\Big(b(p_o^{-1}n)\Big)\,\Big|f\Big(v, v(p_o^{-1}n)v\Big)\Big|^2\,dv\,dn$$

$$= \int_V\int_V \mu\Big(b\Big(p^{-1}p_o v' p^{-1}\Big)\Big)\,\Big|f\Big(v, v\Big(p^{-1}p_o v' p^{-1}\Big)\Big)\Big|^2\,dv\,dv'\ ,$$

(by (4.1.11))

$$= \int_V\int_{V^{p_o^{-1}}} \mu\Big(b\Big(v' p_o^{-1}\Big)\Big)\,\Big|f\Big(v, v\Big(v' p_o^{-1}\Big)\Big)\Big|^2\,dv'\,dv$$

$$= \int_V \int_V |f(v,v')|^2 \, dv' dv \quad ,$$

by Lemma 4.1.9. Similarly

$$\int_V \int_V \left| \left(\Phi_{\lambda_1, \lambda_2} g \right) (v_1, v_2) \right|^2 dv_1 dv_2 = \int_N \int_V |g(n,v)|^2 \, dv \, dn$$

for any $g \in L^2(N \times V)$. Therefore $\Phi_{\lambda_1, \lambda_2}$ is unitary. This completes the proof of Theorem 4.1.10.

Since the tensor product is induced by a character of the Cartan subgroup C we get additional information by applying Theorem 3.5.5 to the special case $C_1 = C$.

THEOREM 4.1.12 *Let G be a connected complex semi-simple Lie group. Suppose $S(\cdot, \lambda_j)$, $j = 1,2,3,4$ are four elements of the non-degenerate principal series. Then*

$$S(\cdot, \lambda_1) \otimes S(\cdot, \lambda_2) \simeq S(\cdot, \lambda_3) \otimes S(\cdot, \lambda_4)$$

if and only if $\lambda_1 \lambda_2 = \lambda_3 \lambda_4$ on the center $\mathbb{Z}(G)$ of G . If we write $\lambda_j = (\xi_j, \eta_j)$, then $\lambda_1 \lambda_2 = \lambda_3 \lambda_4$ on $\mathbb{Z}(G)$ if and only if $\eta_1 + \eta_2 - \eta_3 - \eta_4 \in L_{\mathbb{Z}}$ (see Corollary 3.2.16). *The family of*

<u>representations</u>

$$\{ S(\cdot,\lambda_1)\otimes S(\cdot,\lambda_2) \mid \lambda_1,\lambda_2 \in \hat{C}\}$$

<u>breaks up into exactly</u> n <u>unitary equivalence classes, where</u> n <u>is the order of</u> $\mathbb{Z}(G)$.

<u>Proof</u>: By Theorem 4.1.10

$$S(\cdot,\lambda_i)\otimes S(\cdot,\lambda_j) \simeq \operatorname{ind}_{C\uparrow G} \lambda_i \otimes p_o\lambda_j \quad .$$

Therefore by Theorem 3.5.5

$$S(\cdot,\lambda_1)\otimes S(\cdot,\lambda_2) \simeq S(\cdot,\lambda_3)\otimes S(\cdot,\lambda_4)$$

if and only if $\lambda_1 \otimes p_o\lambda_2 = \lambda_3 \otimes p_o\lambda_4$ on $\mathbb{Z}(G)$. Since $p_o\lambda\Big|_{\mathbb{Z}(G)} = \lambda\Big|_{\mathbb{Z}(G)}$ for any $\lambda \in \hat{C}$ we get $\lambda_1 \otimes p_o\lambda_2 = \lambda_3 \otimes p_o\lambda_4$ on $\mathbb{Z}(G)$. Let $\{\chi_1,\cdots,\chi_n\}$ be the set of unitary characters of $\mathbb{Z}(G)$. If $\lambda_1,\lambda_2 \in \hat{C}$ are arbitrary, then $\lambda_1\lambda_2\Big|_{\mathbb{Z}(G)} = \chi_j$ for some j , $1 \le j \le n$. Hence, since $\lambda_1 \otimes p_o\lambda_2 = \lambda_1\lambda_2$ on $\mathbb{Z}(G)$, it follows that

$$S(\cdot,\lambda_1)\otimes S(\cdot,\lambda_2) \simeq \operatorname{ind}_{C\uparrow G} \lambda_1\lambda_2$$

and

$$\operatorname{ind}_{C\uparrow G} \lambda_1\lambda_2 \simeq \operatorname{ind}_{C\uparrow G} \tilde{\chi}_j$$

by Theorem 3.5.5, where $\tilde{\chi}_j \in C$ is an extension of χ_j to C.

Therefore any $S(\cdot, \lambda_1) \otimes S(\cdot, \lambda_2)$ is unitarily equivalent to one of the representations $\operatorname{ind}_{C \uparrow G} \tilde{\chi}_j$. Conversely $\operatorname{ind}_{C \uparrow G} \tilde{\chi}_j = \operatorname{ind}_{C \uparrow G} 1 \otimes \tilde{\chi}_j$ (where 1 is the trivial character of C) $\simeq S(\cdot, 1) \otimes S(\cdot, \tilde{\chi}_j)$ so that there are exactly n distinct unitary equivalence classes for the representations $S(\cdot, \lambda_1) \otimes S(\cdot, \lambda_2)$, $\lambda_1, \lambda_2 \in \hat{C}$.

4.2 Harish-Chandra's Plancherel formula

We shall focus attention now on the problem of decomposing the tensor product of principal series representations as a direct integral.

A solution of this problem is based on the Mackey Reciprocity Theorem for non-compact groups; this is Theorem 5.1 of [26]. To this extent the results are not as completely explicit as desirable. On the other hand, we shall indicate somewhat explicitly, how the Reciprocity Theorem is obtained for the special case at hand--namely for the pair G , C , where G is a complex semisimple Lie group and C is a Cartan subgroup of G. In fact, in this case, we shall derive the so-called weak form of the Reciprocity Theorem (Theorem 4.1 of [26]) by using the Plancherel formulas for G and C .

We begin by reviewing Harish-Chandra's basis results about harmonic analysis on G; see [22], [23].

Definition 4.2.1 Let L be a bounded linear operator on a separable Hilbert space $\mathcal{H}$. L is said to be traceable (or of the trace class) if for any complete orthonormal system $\{e_j\}_{j=1}^{\infty}$ of $\mathcal{H}$, the series $\sum_{j=1}^{\infty} |(Le_j|e_j)|$ converges. If this is the case the trace of L , denoted by $\operatorname{tr} L$, is $\sum_{j=1}^{\infty} (Le_j|e_j)$.

If $\sum_{j=1}^{\infty} |(Le_j|e_j)|$ converges for a particular orthonormal basis $\{e_j\}_{j=1}^{\infty}$ of $\mathcal{H}$, then L is traceable; moreover $\operatorname{tr} L$ is independent of the choice of basis.

Suppose G_1 is a connected semi-simple Lie group (G_1 is not assumed to be complex). Let $C_c^\infty(G_1)$ be the algebra of indefinitely differentiable, compactly supported functions on G_1. In [22], Harish-Chandra shows that if T is an irreducible unitary representation of G_1, then for every f in $C_c^\infty(G_1)$, the operator

$$\hat{f}(T) = \int_{G_1} f(x)\, T(x)\, dx$$

is traceable and, in fact, the linear functional

$$\Theta_T : f \longrightarrow \operatorname{tr} \hat{f}(T)$$

is a distribution on G_1 in the sense of Schwartz. Moreover Θ_T depends only on the <u>class</u> of unitary representations equivalent to T. Θ_T is called the <u>character</u> of T.

Suppose, in particular, that $G_1 = G$ is complex (some of our remarks are valid more generally). If $T = S(\cdot, \lambda)$, $\lambda \in \hat{B}$, is an element of the (non-degenerate) principal series, we shall write

$$\hat{f}(\lambda) = \int_G f(x)\, S(x, \lambda)\, d\lambda$$

for $f \in L^1(G)$. $\hat{f}(\lambda)$ is the value of the <u>Fourier transform</u> of f at λ. It is known that the character $\Theta_\lambda : f \longrightarrow \operatorname{tr} \hat{f}(\lambda)$, $f \in C_c^\infty(G)$, is actually given by a locally integrable function on G. We set

(4.2.2)
$$I(c, f) = \mu^{\frac{1}{2}}(c) \int_N \int_K f(k^{-1} cnk)\, dk\, dn$$

(4.2.3)
$$\Lambda_f(k, k_1, \lambda) = \int_B f(k^{-1} b\, k_1)\, \lambda(b)\, \mu^{\frac{1}{2}}(b^{-1})\, db\,,$$

$$(c, k, k_1) \in C \times K \times K,\ f \in C_c^{\infty}(G),\ \mu = \frac{1}{\Delta_B}\,.$$

Define

$$L^2(K, \lambda) = \left\{ f \in L^2(K) \mid f(mk) = \lambda(m)\, f(k) \ \text{ for all } \ (m, k) \in M \times K \right\}.$$

Proposition 4.2.4 $S(\cdot, \lambda)$ can be realized as a representation of G on $L^2(K, \lambda)$ as follows:

$$\left(S(x, \lambda)\, f\right)(k) = \mu^{\frac{1}{2}}\left(a(kx)\right)\, \lambda\left(a(kx)\right)\, f\left(k(kx)\right)$$

where

$$G = ANK\,,\ f \in L^2(K, \lambda)\,,\ (k, x) \in K \times G\,.$$

See [19].

Proposition 4.2.4 gives the "compact picture" of $S(\cdot, \lambda)$. On the space $L^2(K, \lambda)$ we have

$$\left(\hat{f}(\lambda)\, \psi\right)(k) = \int_K \Lambda_f(k, k_1, \lambda)\, \psi(k_1)\, dk_1,\ \psi \in L^2(K, \lambda),\ k \in K.$$

Clearly $\Lambda_f(\cdot, \cdot, \lambda)$ is continuous on $K \times K$ so that

$$\int_K \int_K |\Lambda_f(k, k_1, \lambda)|^2\, dk\, dk_1 < \infty\,.$$

Hence $\hat{f}(\lambda)$ is a Hilbert-Schmidt operator and $\Lambda_f(\cdot, \cdot, \lambda)$ is the Hilbert-Schmidt kernel of $\hat{f}(\lambda)$. Also $I(\cdot, f) \in C_c^\infty(G)$ and

$$\operatorname{tr} \hat{f}(\lambda) = \int_K \Lambda_f(k, k, \lambda)\, dk = \int_C I(c, f)\, \lambda(c)\, dc.$$

This shows that for f in $C_c^\infty(G)$, $\operatorname{tr} \hat{f}(\lambda)$ is the Fourier transform of $I(\cdot, f)$ at $\lambda \in \hat{C}$.

Now we come to Harish-Chandra's basic inversion formula:

(4.2.5) $$f(e) = \int_{\hat{C}} \operatorname{tr} \hat{f}(\lambda)\, \omega(\lambda)\, d\lambda$$

for $f \in C_c^\infty(G)$, where $\omega(\lambda)\, d\lambda$ is the Plancherel measure of G; $\omega(\lambda)$ is given by the explicit formula

(4.2.6) $$\omega(\lambda) = a \prod_{\alpha \in \Delta^+} |\xi(H_\alpha) + \eta(H_\alpha)|^2,$$

where $\lambda = (\xi, \eta)$, Δ^+ is the set of positive roots, and $B(H, H_\alpha) = \alpha(H)$ for all $H \in \mathfrak{h}$ (B is the Killing form and $\mathfrak{h}$ is the Cartan subalgebra); a is a positive constant which depends on the normalization of Haar measure $d\lambda$ on $\hat{C}$ (and not on f). The inversion formula leads immediately to the Plancherel formula:

(4.2.7) $$\int_G |f(x)|^2\, dx = \int_{\hat{C}} \operatorname{tr} \hat{f}(\lambda)\, \hat{f}(\lambda)^*\, \omega(\lambda)\, d\lambda$$

for f in $C_c^\infty(G)$ where

$$\operatorname{tr} \hat{f}(\lambda)\, \hat{f}(\lambda)^* = \int_K \int_K |\Lambda_f(k, k_1, \lambda)|^2\, dk\, dk_1 .$$

The operator $\hat{f}(\lambda)$ acting on $L^2(V)$ is, again, an integral operator with kernel $\Lambda_f(\cdot, \lambda)$ given by

(4.2.8) $$\Lambda_f(v, v_1, \lambda) = \int_B f(v^{-1} b v_1)\, \mu^{\frac{1}{2}}(b^{-1})\, \lambda(b)\, db,$$

$(v, v_1) \in V \times V$, $\mu = \frac{1}{\Delta_B}$, $f \in C_c^{\infty}(G)$. If we think of Λ_f as a function on $\hat{C} \times V \times V$, (or on $\hat{C} \times K \times K$) then (4.2.7) implies that the map $F^G : f \longrightarrow \Lambda_f \left(f \in C_c^{\infty}(G)\right)$ is an isometry. F^G extends to a unitary map F^G of $L^2(G)$ onto

$$L^2\left(\hat{C} \times V \times V, \omega(\lambda)\, d\lambda\, dv dv\right)\left(\text{or } L^2\left(\hat{C} \times K \times K, \omega(\lambda)\, d\lambda\, dk dk\right)\right).$$

F^G is called the Plancherel transform of G. Similarly we have the Plancherel transform F^C of the locally compact abelian group C. F^C is a unitary map of $L^2(C)$ onto $L^2(\hat{C})$. We shall use F^G, F^C to analyze the restriction of the two-sided regular representation of $G \times G$ to $C \times G$ and thereby deduce, for the pair G, C, the Mackey Reciprocity Theorem in the weak form.

4.3 The Frobenius-Mackey Reciprocity Theorem for non-compact groups

Let τ^G denote the unitary representation of $G \times G$ on $L^2(G)$ defined by the formula:

$$(4.3.1) \qquad (\tau^G(a_1,a_2)f)(x) = f(a_1^{-1}x\, a_2),$$

$$(a_1,a_2,x) \in G \times G \times G \ , \ f \in L^2(G).$$

We shall call τ^G the <u>two-sided</u> regular representation of $G \times G$.

Proposition 4.3.2 *The transform of* τ^G *under* F^G *acts on the space* $L^2(\hat{C} \times V \times V, \omega(\lambda)\, d\lambda\, dvdv)$ *as follows:*

$$(\tau^G(a_1,a_2)\Lambda)(\lambda,v_1,v_2)$$

$$= \mu^{\frac{1}{2}}(b(v_1\, a_1))\lambda(b(v_1a_1))\mu^{\frac{1}{2}}(b(v_2a_2))\overline{\lambda}(b(v_2a_2))\Lambda(\lambda,v(v_1a_1),v(v_2a_2)),$$

$$\Lambda \in L^2(\hat{C} \times V \times V, \omega(\lambda)d\lambda\, dvdv),\ (a_1,a_2,v_1,v_2,\lambda) \in G \times G \times V \times V \times \hat{C}.$$

In other words τ^G *is decomposed as a direct integral:*

$$\tau^G = \int_{\hat{C}}^{\oplus} S(\cdot,\lambda) \times S(\cdot,\overline{\lambda})\ \omega(\lambda)\, d\lambda\ ,$$

where $S(\cdot,\lambda) \times S(\cdot,\overline{\lambda})$ *is the outer tensor product* (see Appendix).

Proof: For $(a_1,a_2,v_1,v_2,\lambda) \in G \times G \times V \times V \times \hat{C}$ and $f \in C_c^{\infty}(G)$ we have

$$\left(F^G \tau^G(a_1, a_2) F^{G^{-1}} \Lambda_f\right)(\lambda, v_1, v_2) = \Lambda_{\tau(a_1, a_2)f}(\lambda_1, v_1, v_2)$$

$$= \int_B \left(\tau(a_1, a_2)f\right)\left(v_1^{-1} b v_2\right) \mu^{\frac{1}{2}}(b^{-1})\, \lambda(b)\, db \qquad \text{(by 4.2.8)}$$

$$= \int_B f\left(a_1^{-1} v_1^{-1} b v_2 a_2\right) \mu^{\frac{1}{2}}(b^{-1})\, \lambda(b)\, db \qquad \text{(by 4.3.1))}$$

$$= \int_B f\left(v(v_1 a_1)^{-1}\, b(v_1 a_1)^{-1} b\, b(v_2 a_2)\, v(v_2 a_2)\right) \mu^{\frac{1}{2}}(b^{-1})\, \lambda(b)\, db$$

$$= \int_B f\left(v(v_1 a_1)^{-1}\, b(v_1 a_1)^{-1} b\, v(v_2 a_2)\right) \mu^{\frac{1}{2}}\left(b(v_2 a_2)\right) \mu^{\frac{1}{2}}(b^{-1})\, \overline{\lambda}\left(b(v_2 a_2)\right) \lambda(b)\, db$$

$$= \mu^{\frac{1}{2}}\left(b(v_2 a_2)\right) \overline{\lambda}\left(b(v_2 a_2)\right) \int_B f\left(v(v_1 a_1)^{-1} b\, v(v_2 a_2)\right) \Delta_B\left(b(v_1 a_1)^{-1}\right)$$

$$\mu^{-\frac{1}{2}}\left(b(v_1 a_1)\right) \mu^{\frac{1}{2}}(b^{-1})\, \lambda\left(b(v_1 a_1)\right) \lambda(b)\, db$$

$$= \mu^{\frac{1}{2}}\left(b(v_1 a_1)\right) \lambda\left(b(v_1 a_1)\right) \mu^{\frac{1}{2}}\left(b(v_2 a_2)\right) \overline{\lambda}\left(b(v_2 a_2)\right) \Lambda_f\left(\lambda, v(v_1 a_1), v(v_2 a_2)\right)$$

(by (4.2.8)).

Corollary 4.3.3 <u>Let</u> $\tau^{G, C} = \tau^G\big|_{C \times G}$. <u>Then</u>

$$\tau^{G, C} \simeq \int_{\hat{C}}^{\oplus} S(\cdot, \lambda)\big|_C \times S(\cdot, \overline{\lambda})\, \omega(\lambda)\, d\lambda.$$

Next we obtain an alternate decomposition of $\tau^{G, C}$:

Proposition 4.3.4 <u>The equation</u>

$$(\Phi f)(\lambda, n, v) = \int_C f(cnv)\, \overline{\lambda}(c)\, dc,$$

$f \in L^2(G)$, $(\lambda, n, v) \in \hat{C} \times N \times V$, <u>defines a unitary map</u> Φ <u>of</u> $L^2(G)$ <u>onto</u>

$L^2(\widehat{C} \times N \times V,\ d\lambda\, dn\, dv)$. <u>The transform of</u> $\tau^{G,C}$ <u>under</u> Φ <u>acts on</u> $L^2(\widehat{C} \times N \times V,\ d\lambda\, dn\, dv)$ <u>as follows</u>:

$$\left(\tau^{G,C}(c,a)\,\psi\right)(\lambda,n,v) = \lambda(c)\,\overline{\lambda}\left(c(va)\right)\,\psi\left(\lambda, c(va)^{-1} n\, c(va)\, n(va), v(va)\right)$$

$\psi \in L^2(\widehat{C} \times N \times V,\ d\lambda\, dn\, dv)$, $(c,a,n,v,\lambda) \in C \times G \times N \times V \times \widehat{C}$.

<u>In other words</u>

$$\tau^{G,C} = \int_{\widehat{C}}^{\oplus} \lambda \times \operatorname{ind}_{C\uparrow G} \overline{\lambda}\, d\lambda \qquad \text{(see Theorem 3.3.5).}$$

<u>Proof</u>: By the Plancherel Theorem for C,

$$\int_{\widehat{C}} |(\Phi f)(\lambda,n,v)|^2\, d\lambda = \int_C |f(cnv)|^2\, dc$$

for all $(n,v) \in N \times V$. By (3.1.24) we get

$$\int_{\widehat{C}}\int_N\int_V |(\Phi f)(\lambda,n,v)\lambda|^2\, dn\, dv\, d\lambda = \int_N\int_V\int_C |f(cnv)|^2\, dc\, dv\, dn$$

$$= \int_G |f(x)|^2\, dx\ ;$$

hence Φ is an isometry. By standard arguments, Φ extends to a unitary map of $L^2(G)$ onto $L^2(\widehat{C} \times N \times V, d\lambda\, dn\, dv)$. We have

$$\left(\Phi\tau^{G,C}(c,a)\Phi^{-1}\psi\right)(\lambda,n,v) = \int_C \left(\tau^{G,C}(c,a)\Phi^{-1}\psi\right)(c_1 nv)\,\lambda(c_1)\,dc_1$$

$$= \int_C (\Phi^{-1}\psi)(c^{-1}c_1 nva)\,\lambda(c_1)\,dc_1 = \qquad \text{(by (4.3.1))}$$

$$= \lambda(c) \int_C (\Phi^{-1}\psi)(c_1 nva)\, \lambda(c_1)\, dc_1$$

$$= \lambda(c) \int_C (\Phi^{-1}\psi)\Big(c_1\, c(va)\, c(va)^{-1} n\, c(va)\, n(va)\, v(va)\Big)\, \lambda(c_1)\, dc_1$$

$$= \lambda(c)\, \overline{\lambda}\big(c(va)\big) \int_C (\Phi^{-1}\psi)\Big(c_1\, c(va)^{-1} n\, c(va)\, n(va)\, v(va)\Big)\, \lambda(c_1)\, dc_1$$

$$= \lambda(c)\, \overline{\lambda}\big(c(va)\big)\, \psi\Big(\lambda, c(va)^{-1} n\, c(va)\, n(va),\ v(va)\Big).$$

If we combine Corollary 4.3.3 and Proposition 4.3.4 we get

THEOREM 4.3.5 (Frobenius-Mackey Reciprocity Theorem in Weak Form) _Let G be a connected complex semi-simple Lie group and let C be a Cartan subgroup of G. Given a unitary character_ λ _of_ C, _let_ $S(\cdot, \lambda)$ _be the corresponding element of the non-degenerate principal series. Then_

$$\int_{\hat{C}}^{\oplus} \lambda \times \operatorname*{ind}_{C\uparrow G} \overline{\lambda}\, d\lambda \simeq \int_{\hat{C}}^{\oplus} S(\cdot, \lambda)\Big|_C \times S(\cdot, \overline{\lambda})\, \omega(\lambda)\, d\lambda$$

where $\omega(\lambda)$ is the Plancherel density (see (4.2.6)).

Remark: Theorem 4.3.5 is a theorem concerning the specific pair G, C. More generally, Mackey considers an arbitrary (separable) locally compact group G_1 and a closed subgroup H_1 of G_1. If the regular representations of H_1 and G_1 are of _type_ 1, then there is a reciprocity theorem (Theorem 4.1 of [26]).

We shall state the strong form of the Reciprocity Theorem, for the pair G, C, and then comment on its proof.

Prior to this, we consider the <u>decomposition</u> of a finite measure: Suppose (X, ζ) and (Y, ν) are finite Borel measure spaces. Let α be a finite Borel measure on $X \times Y$. If $Q \subset X \times Y$ and $(x_o, y_o) \in X \times Y$, let

$$Q^{x_o} = \{y \in Y \mid (x_o, y) \in Q\}$$

$$Q^{y_o} = \{x \in X \mid (x, y_o) \in Q\} ;$$

Q^{x_o}, Q^{y_o} are the Y and X <u>sections</u> of Q, respectively. By the <u>Measure Decomposition Theorem,</u> [21], there exists for each $(x, y) \in X \times Y$ finite Borel measures α_x, α_y on Y, X, respectively such that

$$\alpha = \int_X \alpha_x \, d\zeta(x) = \int_Y \alpha_y \, d\nu(y)$$

i.e., for every Borel set $Q \subset X \times Y$,

(4.3.6) $$\alpha(Q) = \int_X \alpha_x(Q^x) \, d\zeta(x) = \int_Y \alpha_y(Q^y) \, d\nu(y) .$$

The measures α_x, α_y, $(x, y) \in X \times Y$, are called the x and y <u>slices</u> of α.

THEOREM 4.3.7 (Frobenius-Mackey Reciprocity Theorem in Strong Form) *Let G be a connected complex semi-simple Lie group and let C be a Cartan subgroup of G. Given a unitary character λ of C, let $S(\cdot, \lambda)$ be the corresponding element of the non-degenerate principal series. Suppose ζ, ν are finite Borel measures on $\hat{C}$ equivalent to the Haar measure on $\hat{C}$ and the Plancherel measure on G, respectively ; assume $\zeta(\hat{C}) = \nu(\hat{C})$. Then there exists a finite Borel measure α on $\hat{C} \times \hat{C}$ and a Borel function n from $\hat{C} \times \hat{C}$ to the countable cardinals such that*

$$\text{(i)} \qquad \alpha(E \times \hat{C}) = \zeta(E), \quad \alpha(\hat{C} \times F) = \nu(F)$$

for all Borel sets $E, F \subset \hat{C}$

$$\text{(ii)} \qquad \operatorname*{ind}_{C \uparrow G} \lambda \simeq \int_{\hat{C}}^{\oplus} n(\lambda, \theta)\, S(\cdot, \theta)\, d\alpha_\lambda(\theta)$$

for ζ almost all λ in $\hat{C}$

$$\text{(iii)} \qquad S(\cdot, \theta)\Big|_C \simeq \int_{\hat{C}}^{\oplus} n(\lambda, \theta)\, \lambda d\alpha_\theta(\lambda)$$

for ν almost all θ in $\hat{C}$; α_λ, α_θ are the λ and θ slices of α (see (4.3.6)).

We will show how α is constructed, and later we will show that

$$n(\lambda, \theta) = 1 \text{ or } \infty$$

for all $(\lambda, \theta) \in \hat{C} \times \hat{C}$. In fact we will show that $n \equiv 1$ if and only if the maximal nilpotent subgroup N of G is abelian.

A proof of Theorem 4.3.7 is obtained by applying Mackey's arguments to the present situation. Of course, one makes use of Theorem 4.3.5. Let us indicate how the measure α is obtained: Corresponding to the direct integral decompositions of $\tau^{G,C}$ given by Proposition 4.3.4 and Corollary 4.3.3, we have spectral measures P_1, P_2 on $\hat{C}$ which take values in the set of projections on $L^2(\hat{C} \times N \times V)$, $L^2(\hat{C} \times V \times V)$, respectively. P_1, P_2 are given by

$$P_1(E)\psi = \chi_E \psi$$

$$P_2(F)\Lambda = \chi_F \Lambda$$

where χ_E, χ_F are the characteristic functions of $E, F \subset \hat{C}$, $(\psi, \Lambda) \in L^2(\hat{C} \times N \times V) \times L^2(\hat{C} \times V \times V)$. Clearly P_1 and P_2 commute with $\tau^{G,C}$. Let

$$\tilde{P}_1(\cdot) = \Phi^{-1} P_1(\cdot)\Phi$$

$$\tilde{P}_2(\cdot) = (F^G)^{-1} P_2(\cdot) F^G$$

where F^G is the Plancherel transform on G and Φ is defined by Proposition 4.3.4. Then $\tilde{P}_1$, $\tilde{P}_2$ are spectral measures on $\hat{C}$ with values in the set of projections on $L^2(G)$. One of the crucial steps in the proof of Theorem 4.3.7 is that of showing commutativity of $\tilde{P}_1, \tilde{P}_2$. Mackey shows more; actually the ranges of $\tilde{P}_1$, $\tilde{P}_2$ are contained in the center of the intertwining algebra $\mathcal{R}(\tau^{G,C}, \tau^{G,C})$ of $\tau^{G,C}$. By definition $\mathcal{R}(\tau^{G,C}, \tau^{G,C})$ is the set of bounded operators L on the representation space of $\tau^{G,C}$ such that

$$L\,\tau^{G,C}(c,g) = \tau^{G,C}(c,g)\,L$$

for all $(c,g) \in C \times G$. One can see, directly, that the range of $\tilde{P}_1$ is in the center of $\mathcal{R}(\tau^{G,C}, \tau^{G,C})$ and hence that $\tilde{P}_1$, $\tilde{P}_2$ commute. In fact suppose $L \in \mathcal{R}(\tau^{G,C}, \tau^{G,C})$ is arbitrary. Then for every $(c,g) \in C \times G$, $L\,\tau^{G,C}(c,g) = \tau^{G,C}(c,g)L$; in particular, $L\,\tau^{G,C}(c,e) = \tau^{G,C}(c,e)\,L$. Now identify $L^2(\widehat{C} \times N \times V)$ with

$$L^2\left(\widehat{C}, L^2(N \times V)\right) = \int_{\widehat{C}}^{\oplus} L^2(N \times V)\, d\lambda .$$

Then by Proposition 4.3.4, we have

$$\left(\tau^{G,C}(c,e)\,\psi\right)(\lambda) = \lambda(c)\,\psi(\lambda) \tag{4.3.8}$$

for all $\psi \in L^2\left(\widehat{C}, L^2(N \times V)\right)$ and $c \in C$. As c varies over C, we obtain all characters $\hat{\hat{c}}$ of $\widehat{C}$ by the relation

$$\hat{\hat{c}}(\lambda) = \lambda(c) \qquad \text{(Pontrjagin duality)}$$

$\lambda \in \widehat{C}$. Therefore, by (4.3.8), L commutes with the multiplication operators $M_{\hat{\hat{c}}}$, $c \in C$. By direct integral theory, [2], [29], L is given by an essentially bounded measurable field of operators $B(\lambda)$ over $\widehat{C}$:

$$L = \int_{\hat{C}}^{\oplus} B(\lambda)\, d\lambda .$$

From this it is clear that P_1 and L commute. Therefore $\tilde{P}_1(E)$ is in the center of $\mathcal{R}(\tau^{G,C}, \tau^{G,C})$ for all Borel sets $E \subset \hat{C}$ and, hence, $\tilde{P}_1$ and $\tilde{P}_2$ commute.

Since $\tilde{P}_1$ and $\tilde{P}_2$ commute, there is a unique spectral measure S on $\hat{C} \times \hat{C}$ such that

$$S(E \times F) = \tilde{P}_1(E)\, \tilde{P}_2(F) \tag{4.3.9}$$

for all Borel sets $E, F \subset \hat{C}$; see [25]. The range of S is contained in $\mathcal{R}(\tau^{G,C}, \tau^{G,C})$ and therefore, by [25], S defines a direct integral decomposition of $\tau^{G,C}$ relative to a finite Borel measure α on $\hat{C} \times \hat{C}$. Let ϕ be any element of $L^2(G)$ such that $S(Q)\phi = 0$ if and only if $S(Q) = 0$ for any Q Borel set $Q \subset \hat{C} \times \hat{C}$ (Such a ϕ is known to exist; see [27], page 86.) Then we can define α by setting

$$\alpha(Q) = \big(S(Q)\, \phi \,|\, \phi\big),$$

$Q \subset \hat{C} \times \hat{C}$ a Borel set.

In application, we shall use a more pragmatic version of Theorem 4.3.7. The following Corollary, paraphrased for our purposes, is due to N. Anh ; see [1], Corollary 1.10. We retain the notation of Theorem 4.3.7.

Corollary 4.3.10 *Let* w *and* n' *be Borel functions on* $\hat{C} \times \hat{C}$ *such that* $n'(\lambda, \theta)$ *is a countable cardinal for every* $(\lambda, \theta) \in \hat{C} \times \hat{C}$. *Then the following statements are equivalent:*

a. $$\operatorname*{ind}_{C\uparrow G} \lambda \simeq \int_{\hat{C}}^{\oplus} n'(\lambda, \theta)\, S(\cdot, \theta)\, w(\lambda, \theta)\, d\nu(\theta)$$

for ζ *almost all* λ

b. $$S(\cdot, \theta)\big|_C \simeq \int_{\hat{C}} n'(\lambda, \theta)\, \lambda\, w(\lambda, \theta)\, d\zeta(\lambda)$$

for ν *almost all* θ.

Anh shows, in fact, that a. is equivalent to:

c. $$\begin{cases} w\, d\zeta\, d\nu \text{ is equivalent to } \alpha \\ n' = n \text{ almost everywhere} \end{cases}$$

where α, n are given by Theorem 4.3.7. Similarly b. is equivalent to c.

Since the tensor product of two principal series representations has the form $\operatorname*{ind}_{C\uparrow G} \lambda$ for some $\lambda \in \hat{C}$, see Theorem 4.1.10, Corollary 4.3.10 implies that its decomposition is essentially known once one knows the decomposition of $S(\cdot, \theta)\big|_C$, $\theta \in \hat{C}$. Therefore, we shall now take up the problem of decomposing the restriction of the principal series to the Cartan subgroup.

4.4 The decomposition of the restriction of the principal series to a Cartan subgroup

First we study the (right) action of $C/\mathbb{Z}(G)$ on V :

$$v \cdot \tilde{c} = c^{-1}vc$$

where $\tilde{c} = \mathbb{Z}(G)c$, $(c,v) \in C \times V$. We will throw away a set of measure zero in V . $C/\mathbb{Z}(G)$ will act on the remaining subset of V and the space or orbits will be computed. <u>The non-transitive action of $C/\mathbb{Z}(G)$ will account for the infinite multiplicity of the principal series occurring in the reduction of the tensor product.</u>

Choose an ordering of the positive roots Δ^+ and write

$$\Delta^+ = \{\alpha_1,\ldots,\alpha_\ell,\alpha_{\ell+1},\ldots,\alpha_k\}$$

where $\pi = \{\alpha_1,\ldots,\alpha_\ell\}$ and k is the cardinality of Δ^+ . Fix non-zero elements X_j in $\mathfrak{g}_{-\alpha_j}$, $j = 1,2,\ldots,k$. The map

$$(z_j) \to \prod_{j=1}^{k} \exp z_j X_j \qquad (z_j) \in \mathbb{C}^k$$

is a 1-1 biholomorphic map if $\mathbb{C}^k$ onto V . Thus we can identify Haar measure on V with Lebesgue measure on $\mathbb{C}^k$. If

$$v = \prod_{j=1}^{k} \exp z_j X_j \in V \quad , \quad (z_j) \in \mathbb{C}^k \quad ,$$

we shall write $v = v(z_1,\ldots,z_k)$. We define

$$(4.4.1) \qquad V_o = \{v(z_1,\ldots,z_k)\in V | z_j \neq 0,\ 1 \leq j \leq \ell\}.$$

$$(4.4.2) \qquad V' = \{v(z_1,\ldots,z_k)\in V | z_j = 0,\ 1 \leq j \leq \ell\}.$$

Then $V - V_o$ has measure zero. If $\alpha_j \in \Delta^+$ is not simple we shall write

$$\alpha_j = \sum_{i=1}^{\ell} m_i^{(j)} \alpha_i \tag{4.4.3}$$

where the $m_i^{(j)}$ are non-negative integers. By Theorem 3.2.11, the map

$$c \longrightarrow \left(e^{\alpha_j(H)}\right)_{j=1}^{\ell} \tag{4.4.4}$$

$c = \exp H$, defines an isomorphism of $C/\mathbb{Z}(G)$ onto $\mathbb{C}^{x^{\ell}}$. Hence $\mathbb{C}^{x^{\ell}}$ acts on V. Identify V with $\mathbb{C}^k$; then $\mathbb{C}^{x^{\ell}}$ acts on $\mathbb{C}^k$.

Proposition 4.4.5 $C/\mathbb{Z}(G)$ acts on V_o. If we identify $C/\mathbb{Z}(G)$ with $\mathbb{C}^{x^{\ell}}$ (see (4.4.4)) and identify V_o with $\mathbb{C}^{x^{\ell}} \times \mathbb{C}^{k-\ell}$ then $\mathbb{C}^{x^{\ell}}$ acts on $\mathbb{C}^{x^{\ell}} \times \mathbb{C}^{k-\ell}$ as follows:

$$(w_1, \cdots, w_\ell)(z_1, \cdots, z_\ell, z_{\ell+1}, \cdots, z_k)$$
$$= \left(w_1 z_1, \cdots, w_\ell z_\ell, \prod_{j=1}^{\ell} w_j^{m_j^{(\ell+1)}} z_{\ell+1}, \cdots, \prod_{j=1}^{\ell} w_j^{m_j^{(k)}} z_k \right) \text{ (see 4.4.3).}$$

Let $v_o = \prod_{j=1}^{\ell} \exp X_j \in V$. If $\ell = k$ the action of $C/\mathbb{Z}(G)$ on V_o is transitive. If $\ell < k$ the map

$$i : (z_{\ell+1}, \cdots, z_k) \longrightarrow \text{orbit of } \left(v_o \prod_{j=\ell+1}^{k} \exp z_j X_j \right)$$

<u>identifies the space of orbits with</u> $\mathbb{C}^{k-\ell} = V'$.

<u>Proof</u>: Let $v = v(z_1, z_2, \cdots, z_k) \in V_o$ be arbitrary. Then $(z_1, z_2, \cdots, z_\ell) \in \mathbb{C}^{x^\ell}$, so by (4.4.4) there exists $c \in C$ such that

(4.4.6) $$e^{\alpha_j(H)} = z_j$$

$j = 1, 2, \cdots, \ell$, $c^{-1} = \exp H$. Then

$$c^{-1} v c = c^{-1} \exp z_1 X_1 c\, c^{-1} \exp z_2 X_2 c \cdots c^{-1} \exp z_k X_k c$$

$$= \exp z_1 e^{-\alpha_1(H)} X_1 \cdots \exp z_k e^{-\alpha_k(H)} X_k \quad \text{(by Proposition 3.2.8)}$$

$$= v_o \prod_{j=\ell+1}^{k} \exp z_j e^{-\alpha_j(H)} X_j \quad \text{(by (4.4.6))}.$$

For $\ell + 1 \le j \le k$, α_j is non-simple so

$$\alpha_j = \sum_{i=1}^{\ell} m_i^{(j)} \alpha_i$$

by (4.4.3). Hence

$$e^{-\alpha_j(H)} = \prod_{i=1}^{\ell} z_i^{-m_i(j)} \quad \text{(by (4.4.6))},$$

i.e., for any $v = v(z_1, \cdots, z_k) \in V_o$ there exist $c \in C$ such that

(4.4.7) $$c^{-1} v c = v_o \prod_{j=\ell+1}^{k} \exp z_j \prod_{i=1}^{\ell} z_i^{-m_i(j)} X_j .$$

By similar arguments the first statement of Proposition 4.4.5 is easily established. By (4.4.7) the map i is onto. It is also 1-1. For suppose

$$i(z_{\ell+1}, \cdots, z_k) = i(u_{\ell+1}, \cdots, u_k) .$$

Then there exists c in C such that

$$c^{-1} v_o \prod_{j=\ell+1}^{k} \exp z_j X_j \, c = v_o \prod_{j=\ell+1}^{k} \exp u_j X_j . \tag{4.4.8}$$

If $c^{-1} = \exp H$, then (4.4.8) implies that

$$\prod_{j=1}^{\ell} \exp e^{-\alpha_j(H)} X_j \prod_{j=\ell+1}^{k} \exp e^{-\alpha_j(H)} z_j X_j = \prod_{j=1}^{\ell} \exp X_j \prod_{j=\ell+1}^{k} \exp u_j X_j .$$

Therefore $e^{-\alpha_j(H)} = 1$ for $i = 1,2, \cdots, \ell$ and $e^{-\alpha_j(H)} z_j = u_j$, $j = \ell+1, \cdots, k$. For

$$\ell+1 \le j \le k, \ \alpha_j = \sum_{i=1}^{\ell} m_i^{(j)} \alpha_i \qquad ((4.4.3))$$

so that

$$e^{-\alpha_j(H)} = \prod_{i=1}^{\ell} \left[e^{-\alpha_i(H)} \right]^{m_i^{(j)}} = 1 ;$$

hence

$$z_j = u_j , \ \ell+1 \le j \le k.$$

We shall need the following formula which reduces integration over V_o to integration over $C/\mathbb{Z}(G)$ and the space of orbits.

Proposition 4.4.9 Let v_o be as in Proposition 4.4.5. If $f \in C_c(V_o)$ then

$$\int_{V_o} f(v)\, dv = \int_{V'} \int_{C/\mathbb{Z}(G)} \mu(c)\, f(c^{-1} v_o\, v' c)\, d\tilde{c}\, dv'$$

where $\mu = \Delta_B^{-1}$, dv, $d\tilde{c}$ are Haar measures on V, $C/\mathbb{Z}(G)$ and dv' is Lebesgue measure on V' (cf. Proposition 3.6.4).

Proof: We shall use the formula

$$\mu(c) = \exp\left[2\sigma(\mathrm{Re}\log c)\right] \tag{4.4.10}$$

where

$$\sigma = \sum_{\alpha\in\Delta^+} \alpha \qquad \text{(see (3.1.25))}.$$

Let

$$I = \int_{V'} \int_{C/\mathbb{Z}(G)} \mu(c)\, f(c^{-1} v_o\, v' c)\, d\tilde{c}\, dv' .$$

Then

$$I = \int_{\mathbb{C}^{k-\ell}} \int_{\mathbb{C}^{x\,\ell}} \mu\left((w_1, \cdots, w_\ell)\right) f\left(\prod_{j=1}^{\ell} \exp w_j X_j \prod_{j=\ell+1}^{k} \left(\prod_{i=1}^{\ell} w_i^{m_i^{(j)}} \right) z_j X_j \right) \prod_{i=1}^{\ell} \frac{dw_i}{|w_i|^2} \prod_{j=\ell+1}^{k} dz$$

(by Proposition 4.4.5) where $c = \exp H$ and $e^{\alpha_j(H)} = w_j$, $j = 1,2,\cdots,\ell$. Then $e^{\operatorname{Re}\alpha_j(H)} = |w_j|$ so that by (4.4.10)

$$\mu\big((w_1,\cdots,w_\ell)\big) = \prod_{j=1}^{\ell} |w_j|^2 \prod_{j=\ell+1}^{k} e^{2\operatorname{Re}\alpha_j(H)}$$

$$= \prod_{j=1}^{\ell} |w_j|^2 \prod_{j=\ell+1}^{k} \left(\prod_{i=1}^{\ell} |w_i|^{2m_i^{(j)}} \right) \qquad \text{(by (4.4.3))}$$

$$(4.4.11) \qquad = \prod_{j=1}^{\ell} |w_j|^2 \prod_{i=1}^{\ell} |w_i|^{2\sum_{j=\ell+1}^{k} m_i^{(j)}} .$$

Now make the change of variables

$$u_j = w_j \qquad j = 1,2,\cdots,\ell$$

$$u_j = \prod_{i=1}^{\ell} w_i^{m_i^{(j)}} z_j, \quad j = \ell+1,\cdots,k .$$

The square of the modulus of the determinant of the Jacobian matrix is

$$\left| \prod_{j=\ell+1}^{k} \left(\prod_{i=1}^{\ell} w_i^{-m_i^{(j)}} \right) \right|^2 = \prod_{i=1}^{\ell} |w_i|^{-2\sum_{j=\ell+1}^{k} m_i^{(j)}} .$$

Therefore by (4.4.11) we get

$$I = \int_{\mathbb{C}^{k-\ell}} \int_{\mathbb{C}^{\ell}} f\left(\prod_{j=1}^{k} u_j X_j \right) \prod_{j=1}^{k} du_j = \int_{V_o} f(v)\, dv .$$

Corollary 4.4.12 *The equation*

$$(Tf)(\beta, v') = \int_{C/\mathbb{Z}(G)} \mu^{\frac{1}{2}}(c)\, f(c^{-1} v_o v' c)\, \overline{\beta}(\tilde{c})\, d\tilde{c}$$

$f \in L^2(V)$, *defines a unitary map* T *of* $L^2(V)$ *onto* $L^2(\widehat{C/\mathbb{Z}(G)} \times V')$.

Proof: The proof follows immediately from Proposition 4.4.9 and the Plancherel formula for the abelian group $C/\mathbb{Z}(G)$.

THEOREM 4.4.13 *Let* G *be a connected complex semi-simple Lie group and let* C *be a Cartan subgroup of* G. *If* λ *is a unitary character of* C, *let* $S(\cdot, \lambda)$ *be the corresponding element of the* (non-degenerate) *principal series. Let* T *be the unitary map of* $L^2(V)$ *onto* $L^2\left(\widehat{C/\mathbb{Z}(G)} \times V'\right)$ *defined by*

$$(Tf)(\beta, v') = \int_{C/\mathbb{Z}(G)} \mu^{\frac{1}{2}}(c)\, f(c^{-1} v_o v' c)\, \overline{\beta}(\tilde{c})\, d\tilde{c}$$

(see Corollary 4.4.12, (4.4.2), and Proposition 4.4.5). *Let* k *be the number of positive roots and let* ℓ *be the number of simple roots. Then the transform of* $S(\cdot, \lambda)|_C$ *by* T *acts on the space*

$$L^2\left(\widehat{C/\mathbb{Z}(G)} \times V'\right) = L^2\left(\widehat{C/\mathbb{Z}(G)} \times \mathbb{C}^{k-\ell}\right)$$

as follows:

$$\left(S(c, \lambda)f\right)(\beta, v') = \lambda(c)\, \beta(\tilde{c})\, f(\beta, v')$$

$f \in L^2(\widehat{C/\mathbb{Z}(G)} \times V')$, $\tilde{c} = \mathbb{Z}(G)c$. Identify $\widehat{C/\mathbb{Z}(G)}$ with $\mathbb{Z}(G)^{\perp}$.

Then

$$S(\cdot, \lambda)\Big|_C = \int_{\mathbb{Z}(G)^{\perp}}^{\oplus} \epsilon \cdot \lambda\beta \, d\beta^{\perp}$$

where $d\beta^{\perp}$ is Haar measure on $\mathbb{Z}(G)^{\perp}$; i.e., $S(\cdot, \lambda)\Big|_C$ is a direct integral of the unitary characters of C which agree with λ on $\mathbb{Z}(G)$. The multiplicity $\epsilon(\lambda\beta)$ of $\lambda\beta$ in the decomposition is either 1 or ∞ according as $k = \ell$ or $k > \ell$, i.e. according as the maximal nilpotent subgroup N is abelian or non-abelian.

Proof: Let $f \in L^2(\widehat{C/\mathbb{Z}(G)} \times V')$ and let $c \in C$. Then by Theorem 4.1.3

$$\left(T\, S(c, \lambda)\, T^{-1} f\right)(\beta, v') = \int_{C/\mathbb{Z}(G)} \mu^{\frac{1}{2}}(c_1)\left(S(c, \lambda) T^{-1} f\right)\left(c_1^{-1} v_o v' c_1\right) \overline{\beta}(\tilde{c}_1)\, d\tilde{c}_1$$

$$= \int_{C/\mathbb{Z}(G)} \mu^{\frac{1}{2}}(c_1)\, \mu^{\frac{1}{2}}(c)\, \lambda(c)\, (T^{-1} f)\left(c^{-1} c_1^{-1} v_o v' c_1 c\right) \overline{\beta}(\tilde{c}_1)\, d\tilde{c}_1$$

$$= \mu^{\frac{1}{2}}(c)\, \lambda(c) \int_{C/\mathbb{Z}(G)} \mu^{\frac{1}{2}}(c_1) \mu^{\frac{1}{2}}(c^{-1})\, T^{-1} f\left(c_1^{-1} v_o v' c_1\right) \beta(\tilde{c}) \overline{\beta}(\tilde{c}_1)\, d\tilde{c}_1$$

$$= \lambda(c)\, \beta(\tilde{c})\, f(\beta, v') .$$

Clearly

$$\dim L^2(\mathbb{C}^{k-\ell}) = 0 \text{ or } \infty$$

according as $k = \ell$ or $k > \ell$. It is easy to check that $k = \ell$ if and only if $\sum_{\alpha\in\Delta^+} \mathfrak{g}_\alpha$ is abelian; i.e. if and only if N is abelian; see 3.1.6.

4.5 The decomposition of the tensor product of principal series representations

In this final section we decompose the tensor product of principal series representations. If $\lambda \in \hat{C}$ we define

(4.5.1) $$\hat{C}_\lambda = \{\chi \in \hat{C} \mid \chi = \lambda \text{ on } \mathbb{Z}(G)\}.$$

Note that $\hat{C}_\lambda = \lambda\, \mathbb{Z}(G)^\perp$ is a closed subspace of $\hat{C}$ homeomorphic to $\mathbb{Z}(G)^\perp$. If $dz^\perp$ is Haar measure on $\mathbb{Z}(G)^\perp$, then we can define a unique measure $dz_\lambda{}^\perp$ on $\hat{C}_\lambda$ such that

(4.5.2) $$\int_{\hat{C}_\lambda} f(\chi)\, dz_\lambda{}^\perp = \int_{\mathbb{Z}(G)^\perp} f(\lambda\chi)\, dz^\perp(\chi)$$

for every $f \in C_c(\hat{C}_\lambda)$. Normalize Haar measure $d\tilde{\chi}$ on $\hat{C}/\mathbb{Z}(G)^\perp$ so that

(4.5.3) $$\int_{\hat{C}} f(\theta)\, d\theta = \int_{\hat{C}/\mathbb{Z}(G)^\perp} \int_{\mathbb{Z}(G)^\perp} f(\gamma\chi)\, dz^\perp(\gamma)\, d\tilde{\chi}$$

for every $f \in C_c(\hat{C})$. Since $\hat{C}_\lambda$ is a closed subspace of $\hat{C}$ the measure $dz_\lambda{}^\perp$ extends to a unique measure dz_λ on $\hat{C}$ such that

(4.5.) $$\int_{\hat{C}} f\, dz_\lambda = \int_{\hat{C}_\lambda} f|_{\hat{C}_\lambda}\, dz_\lambda{}^\perp$$

for every $f \in C_c(\hat{C})$; see [16]. Moreover

$$(4.5.4) \qquad \int_{\hat{C}-\hat{C}_\lambda} dz_\lambda = 0 .$$

The spaces $L^2(\hat{C}, dz_\lambda)$, $L^2(\hat{C}_\lambda, dz_\lambda^{\perp})$ are unitarily equivalent. On the other hand, the equation

$$(T_\lambda f)(\lambda\chi) = f(\chi) ,$$

$f \in L^2\left(\mathbb{Z}(G)^{\perp}, dz^{\perp}\right)$, defines a unitary map T_λ of $L^2\left(\mathbb{Z}(G)^{\perp}, dz^{\perp}\right)$ onto $L^2(\hat{C}_\lambda, dz_\lambda^{\perp})$, by (4.5.2). Hence there is a unitary map

$$\tilde{T}_\lambda : L^2\left(\mathbb{Z}(G)^{\perp} \times V', dz^{\perp} dv'\right) \longrightarrow L^2(\hat{C} \times V', dz_\lambda dv')$$

such that

$$(\tilde{T}_\lambda f)(\lambda\chi, v') = f(\chi, v')$$

for all $(\chi, v') \in \mathbb{Z}(G)^{\perp} \times V'$. We easily get

Proposition 4.5.5 <u>The transform of</u> $S(\cdot, \lambda)\big|_C$ <u>under</u> $\tilde{T}_\lambda$ <u>acts on the space</u> $L^2(\hat{C} \times V', dz_\lambda dv')$ <u>as follows</u>:

$$\left(S(c, \lambda) f\right)(\lambda\chi, v') = (\lambda\chi)(c) f(\lambda\chi, v')$$

<u>for</u>

$$f \in L^2(\hat{C} \times V', dz_\lambda dv'), \; (c, \chi, v') \in C \times \mathbb{Z}(G)^{\perp} \times V'.$$

<u>Since</u> $\hat{C} - \hat{C}_\lambda$ <u>has</u> dz_λ <u>measure zero</u> (by 4.5.4) <u>we have</u>

$$S(\cdot, \lambda)\big|_C = \int_{\hat{C}}^{\oplus} \varepsilon \cdot \theta \, dz_\lambda(\theta)$$

where $\epsilon = 1$ *or* ∞ *according as* N *is abelian or non-abelian.*

Define $w_o : \hat{C} \times \hat{C} \to \{0, 1\}$ by

$$(4.5.6) \qquad w_o(\gamma, \theta) = \begin{cases} 1 & \text{if } \gamma = \theta \text{ on } \mathbb{Z}(G) \\ 0 & \text{if } \gamma \neq \theta \text{ on } \mathbb{Z}(G) \end{cases} .$$

w_o is clearly a Borel function.

Proposition 4.5.7 *Let* $\lambda \in \hat{C}$ *and let* $d\theta$ *denote Haar measure on* $\hat{C}$. *Then for every* f *in* $C_c(\hat{C})$ *we have*

$$\int_{\hat{C}} f(\theta)\, w_o(\theta, \lambda)\, d\theta = \frac{1}{n} \int_{\hat{C}} f\, dz_\lambda$$

where n *is the order of* $\mathbb{Z}(G)$. *Therefore*

$$n\, w_o(\cdot, \lambda)\, d\theta = dz_\lambda .$$

Proof: Let $\{\chi_1, \cdots, \chi_n\}$ be the set of characters of $\mathbb{Z}(G)$. Extend each χ_j to a unitary character χ_j of C. Since $\hat{C}/\mathbb{Z}(G)^\perp$ is a finite group $\left(\text{isomorphic to } \widehat{\mathbb{Z}(G)}\right)$ we have, by (4.5.3)

$$\int_{\hat{C}} f(\theta)\, w_o(\theta, \lambda)\, d\theta = \frac{1}{n} \sum_{j=1}^{n} \int_{\mathbb{Z}(G)^\perp} f(\gamma\chi_j)\, w_o(\gamma\chi_j, \lambda)\, dz^\perp(\gamma).$$

For some i, $\lambda = \chi_i$ on $\mathbb{Z}(G)$. If $\gamma \in \mathbb{Z}(G)^\perp$ is arbitrary, then $w_o(\gamma\chi_j, \lambda) = \delta_{ij}$ by (4.5.6). Therefore

$$\int_{\hat{C}} f(\theta)\, w_o(\theta, \lambda)\, d\theta = \frac{1}{n} \int_{\mathbb{Z}(G)^{\perp}} f(\gamma \chi_i)\, dz^{\perp}(\gamma)$$

$$= \frac{1}{n} \int_{\hat{C}_{\chi_i}} f(\chi)\, dz_{\chi_i}^{\perp} \qquad \text{(by (4.5.2))}.$$

However $\hat{C}_{\chi_i} = \hat{C}_{\lambda}$ so that $dz_{\chi_i}^{\perp} = dz_{\lambda}^{\perp}$. Also

$$\int_{\hat{C}_{\lambda}} f(\chi)\, dz_{\lambda}^{\perp} = \int_{\hat{C}} f\, dz_{\lambda} \qquad \text{(by (4.5.3))}.$$

Propositions 4.5.5 and 4.5.7 imply that for every $\lambda \in \hat{C}$,

$$S(\cdot, \lambda)\Big|_C = \int_{\hat{C}}^{\oplus} \epsilon \cdot \theta\, n w_o(\theta, \lambda)\, d\theta\ .$$

In Corollary 4.3.10 take $n' = \epsilon$ and $w = n w_o$. Since ζ and ν are equivalent to Haar measure $d\theta$ on $\hat{C}$ and the Plancherel measure $\omega(\theta)\, d\theta$ on G, respectively, we get (by Corollary 4.3.10)

$$\underset{C\uparrow G}{\operatorname{ind}}\ \theta \simeq \int_{\hat{C}}^{\oplus} \epsilon \cdot S(\cdot, \lambda)\, n w_o(\theta, \lambda)\, \omega(\lambda)\, d\lambda$$

for $d\theta$ almost all $\theta \in \hat{C}$. Again $n w_o(\theta, \cdot)\, d\lambda = dz_{\theta}$ and since $\int_{\hat{C}-C_{\theta}} dz_{\theta} = 0$ (by (4.5.4)), we have

(4.5.8)
$$\underset{C\uparrow G}{\operatorname{ind}}\ \theta \simeq \int_{\hat{C}_{\theta}}^{\oplus} \epsilon \cdot S(\cdot, \lambda)\, \omega(\lambda)\, dz_{\theta}^{\perp}$$

for $d\theta$ almost all $\theta \in \hat{C}$. Now (4.5.8) must be true <u>for every</u> θ in $\hat{C}$. To see this let

$$N_\theta = \left\{\theta \in \hat{C} \,|\, (4.5.8) \text{ is false}\right\}.$$

If $\theta_o \in N_\theta$, then $\hat{C}_{\theta_o} \subset N_\theta$ by Theorem 3.5.5 so that $\hat{C}_{\theta_o}$ has Haar measure zero. Let $\{\chi_1, \cdots, \chi_n\}$ be the set of characters of $\mathbf{Z}(G)$ and extend each χ_j to $\chi_j \in \hat{C}$. Then $\hat{C}$ is the disjoint union of the closed sets $\hat{C}_{\chi_j}$, $j = 1, 2, \cdots, n$. Let $\theta_o = \chi_i$ on $\mathbf{Z}(G)$; then $\hat{C}_{\chi_i} = \hat{C}_{\theta_o}$. Also

$$\hat{C} - \hat{C}_{\theta_o} = \bigcup_{j \neq i} \hat{C}_{\chi_j}$$

which shows that $\hat{C}_{\theta_o}$ is open. Since $\hat{C}_{\theta_o} \neq \emptyset$, $\hat{C}_{\theta_o}$ cannot have zero Haar measure.

By virtue of Theorem 4.1.10, we have proved, in summary,

THEOREM 4.5.9 <u>Let</u> G <u>be a connected complex semi-simple Lie group and let</u> C <u>be a Cartan subgroup of</u> G. <u>Suppose that</u> $S(\cdot, \lambda_1)$ <u>and</u> $S(\cdot, \lambda_2)$ <u>are two non-degenerate principal series elements corresponding to the unitary characters</u> λ_1, λ_2 <u>of</u> $\hat{C}$. <u>Let</u> $\mathbf{Z}(G)$ <u>be the center of</u> G <u>and let</u> $dz^\perp$ <u>be Haar measure on the annihilator of</u> $\mathbf{Z}(G)$. <u>Then the tensor product</u> $S(\cdot, \lambda_1) \otimes S(\cdot, \lambda_2)$ <u>can be decomposed as a direct integral of elements from the principal series:</u>

$$S(\cdot, \lambda_1) \otimes S(\cdot, \lambda_2) \simeq \int^{\oplus}_{\mathbf{Z}(G)^\perp} \epsilon \cdot S(\cdot, \lambda\, \lambda_1 \lambda_2)\, \omega(\lambda)\, dz^\perp(\lambda).$$

<u>The elements</u> $S(\cdot, \gamma)$ <u>occurring in the decomposition are precisely</u>

those for which γ *and* $\lambda_1\lambda_2$ *coincide on* $\mathscr{Z}(G)$. $\Big(\omega(\lambda)$ *is the Plancherel density on* $G\Big)$. *The multiplicity* ε *of each* $S(\cdot, \gamma)$ *in the decomposition is either* 1 *or* ∞, *independently of* γ. $\varepsilon = 1$ *if and only if the maximal nilpotent subgroup* N *of* G, *generated by the positive roots, is abelian.*

APPENDIX

The Frobenius-Mackey Reciprocity Theorem (in weak form) for Compact Groups

We shall justify, following Mackey, the labeling of Theorem 4.3.5 as a "reciprocity" theorem. Indeed, in the compact case a statement of unitary equivalence similar to that given in Theorem 4.3.5 (with direct integration replaced by direct summation of course) is entirely equivalent to the classical Frobenius Reciprocity Theorem; see [26].

If T and S are representations of locally compact groups C_1 and G_1 then we shall denote the outer tensor product of T and S by $T \times S$. Thus $T \times S$ is the representation of the Cartesian product $C_1 \times G_1$ defined by

$$(T \times S)(c,a) = T(c) \otimes S(a)$$

where $(c,a) \in C_1 \times G_1$. One knows that $T \times S$ is irreducible if and only if T and S are irreducible.

Suppose $C_1 = G_1$ in particular. Then the restriction of $T \times S$ to the diagonal

$$G_1^{\Delta} = \{(a,a) | a \in G_1\}$$

gives a representation $T \otimes S$ of G which is, by definition, the <u>inner</u> tensor product (or more simply the tensor product) of T and S .

Theorem 4.3.5 is roughly a statement of the following sort:

$$\text{(A.1)} \qquad \int_{\hat{C}_1}^{\oplus} \beta \times \operatorname*{ind}_{C_1 \uparrow G_1} \overline{\beta}\, d\beta \simeq \int_{\hat{G}_1}^{\oplus} \gamma\big|_{C_1} \times \overline{\gamma}\, d\gamma$$

where G_1 is a non-compact group, C_1 is a closed subgroup of G_1, β , γ are irreducible representations of C_1 , G_1 appearing in the decomposition of the regular representations of C_1 , G_1 respectively, and where the bar denotes the <u>contragredient</u> representation, at least when β is finite-dimensional .

$$\overline{\beta}(c) = \beta(c^{-1})^{\text{transpose}} \qquad c \in C_1 .$$

The point is to show that a statement like (A.1), made precise, in the compact case is equivalent to the Frobenius Reciprocity Theorem. Therefore, we now suppose that G_1 is a compact group and C_1 is a closed subgroup of G_1 . Let $\hat{G}_1$, $\hat{C}_1$ denote the dual spaces of irreducible unitary representations of G_1 , C_1 respectively. Elements in $\hat{G}_1$, $\hat{C}_1$ are necessarily finite-dimensional. We have

Theorem A.2 (Classical Frobenius Reciprocity Theorem) *If* $\beta \in \hat{C}_1$ *and* $\gamma \in \hat{G}_1$ *, then the multiplicity of* β *in* $\gamma|_{C_1}$ *equals the multiplicity of* γ *in* $\operatorname{ind}_{C_1 \uparrow G_1} \beta$.

We shall show that Theorem A.2 is equivalent to the following

Theorem A.3 (Frobenius-Mackey Reciprocity Theorem in Weak Form for Compact Groups) *For* C_1 , G_1 *as above*

$$\sum_{\beta \in \hat{C}_1}^{\oplus} \overline{\beta} \times \operatorname{ind} \beta \simeq \sum_{\gamma \in \hat{G}_1}^{\oplus} \overline{\gamma}|_{C_1} \times \gamma$$

where the bar denotes the contragredient representation.

The next result will be needed.

Proposition A.4 *Let* 1_{G_1} *denote the 1-dimensional trivial representation of* G_1 . *If* T , $S \in \hat{G}_1$, *then* 1_{G_1} *is contained in* $T \otimes S$ *if and only if*

$$T \simeq \overline{S} ;$$

hence 1_{G_1} *is contained in* $T \otimes \overline{S}$ *at most once.*

Proposition A.4 is well-known of course. A short proof of it, which uses character theory, goes as follows:

The character $\chi_{T\otimes S}$ of $T \otimes S$ is $\chi_T \chi_S$ and the character $\chi_{\overline{S}}$ of $\overline{S}$ is $\overline{\chi}_S$, where χ_T, χ_S are the characters of T, S respectively. Therefore the multiplicity of 1_{G_1} in $T \otimes S$ is

$$\int_{G_1} \chi_{T\otimes S}\, da = \int_{G_1} \chi_T \chi_S \, da$$

$$= \int_{G_1} \chi_T \overline{\overline{\chi}}_S \, da$$

$$= \left.\begin{matrix} 1 \text{ if and only if } T \simeq \overline{S} \\ 0 \text{ if and only if } T \simeq \overline{S} \end{matrix}\right\} .$$

Now we show that Theorem A.4 implies Theorem A.3. For $(\beta, \gamma) \in \hat{C}_1 \times \hat{G}_1$ let $n(\gamma, \operatorname{ind}_{C_1\uparrow G_1} \beta)$ denote the multiplicity of γ in $\operatorname{ind}_{C_1\uparrow G_1} \beta$ and, similarly, let $n(\beta, \gamma|_{C_1})$ denote the multiplicity of β in $\gamma|_{C_1}$. By complete reducibility, we can write

$$\operatorname{ind}_{C_1\uparrow G_1} \beta = \sum_{\gamma\in\hat{G}_1}^{\oplus} n(\gamma, \operatorname{ind}_{C_1\uparrow G_1} \beta)\gamma$$

$$\gamma|_{C_1} = \sum_{\beta\in\hat{C}_1}^{\oplus} n(\beta, \gamma|_{C_1})\beta$$

By Theorem A.3

$$\text{(A.6)} \qquad \sum_{\beta\in\hat{C}_1}^{\oplus} \bar{\beta} \times \operatorname{ind} \beta \simeq \sum_{\gamma\in\hat{G}_1}^{\oplus} \bar{\gamma}\big|_{C_1} \times \gamma .$$

(A.5) and (A.6) imply

$$\text{(A.7)} \qquad \sum_{(\beta,\gamma)\in\hat{C}_1\times\hat{G}_1}^{\oplus} n(\gamma, \operatorname{ind}_{C_1\uparrow G_1} \beta)\, \beta \times \gamma \simeq \sum_{(\beta,\gamma)\in\hat{C}_1\times\hat{G}_1}^{\oplus} n(\beta, \gamma\big|_{C_1})\, \beta \times \gamma .$$

By the uniqueness of the decomposition of a representation into irreducible components, we deduce from (A.7) that

$$n(\gamma, \operatorname{ind}_{C_1\uparrow G_1} \beta) = n(\beta, \gamma\big|_{C_1})$$

for all (β, γ) in $\hat{C}_1 \times \hat{G}_1$. This proves Theorem A.2.

Next assume Theorem A.2. Then we can prove

Proposition A.8 *Let* G_1^{Δ} *be the diagonal of* $G_1 \times G_1$ *and let* $1_{G_1^{\Delta}}$ *denote the 1-dimensional trivial representation of* G_1^{Δ} . *Then*

$$\operatorname{ind}_{G_1^{\Delta}\uparrow G_1\times G_1} 1_{G_1^{\Delta}} \simeq \sum_{\gamma\in\hat{G}_1}^{\oplus} \bar{\gamma} \times \gamma .$$

Proof: By complete reducibility

$$\operatorname{ind}_{G_1^\Delta \uparrow G_1\times G_1} 1_{G_1^\Delta} = \sum_\alpha n(T_\alpha \times S_\alpha, \operatorname{ind}_{G_1^\Delta \uparrow G_1\times G_1} 1_{G_1^\Delta})\, T_\alpha \times S_\alpha$$

where T_α , S_α vary over the dual $\hat{G}_1$ of G_1 . The question is which $T_\alpha \times S_\alpha$ actually occur in this decomposition? By Theorem A.2

$$n(T_\alpha \times S_\alpha, \operatorname{ind}_{G_1^\Delta \uparrow G_1\times G_1} 1_{G_1^\Delta}) = n(1_{G_1^\Delta}, (T_\alpha \times S_\alpha)\big|_{G_1^\Delta}) .$$

But

$$n(1_{G_1^\Delta}, (T_\alpha \times S_\alpha)\big|_{G_1^\Delta}) = n(1_G, T_\alpha \otimes S_\alpha)$$

and

$$n(1_G, T_\alpha \otimes S_\alpha) = \left.\begin{matrix} 1 \text{ if and only if } T_\alpha \simeq S_\alpha \\ \\ 0 \text{ if and only if } T_\alpha \not\simeq S_\alpha \end{matrix}\right\}$$

by Proposition A.4. Therefore

$$\operatorname{ind}_{G_1^\Delta \uparrow G_1\times G_1} \simeq \sum_\alpha^\oplus \bar{S}_\alpha \times S_\alpha$$

which proves Proposition A.8.

Define

(A.9) $$\zeta = \operatorname{ind}_{G_1^{\Delta}\uparrow G_1 \times G_1} 1_{G_1^{\Delta}} \Big|_{C_1 \times G_1} .$$

Then

(A.10) $$\zeta \simeq \sum_{\gamma \in \hat{G}_1}^{\oplus} \bar{\gamma}\Big|_{C_1} \times \gamma$$

by Proposition A.8. On the other hand

$$G_1 \times G_1 = G_1^{\Delta}(C_1 \times G_1)$$

and

$$G_1^{\Delta} \cap (C_1 \times G_1) = C_1^{\Delta} .$$

Therefore

(A.11) $$\zeta \simeq \operatorname{ind}_{C_1^{\Delta}\uparrow C_1 \times G_1} 1_{C_1^{\Delta}} ;$$

(A.11) follows directly or by a trivial application of the subgroup theorem. Inducing is stages, we have

$$\zeta \simeq \operatorname{ind}_{C_1\times C_1\uparrow C_1\times G_1} \operatorname{ind}_{C_1^{\Delta}\uparrow C_1\times C_1} 1_{C_1^{\Delta}}$$

by (A.11)

$$\simeq \operatorname{ind}_{C_1\times C_1\uparrow C_1\times G_1} \sum_{\beta\in\hat{C}_1}^{\oplus} \overline{\beta}\times\beta$$

by Proposition A.8

$$\simeq \sum_{\beta\in\hat{C}_1}^{\oplus} \operatorname{ind}_{C_1\times C_1\uparrow C_1\times G_1} \overline{\beta}\times\beta$$

$$\simeq \sum_{\beta\in\hat{C}_1}^{\oplus} \operatorname{ind}_{C_1\uparrow C_1} \overline{\beta}\times \operatorname{ind}_{C_1\uparrow G_1} \beta$$

$$\simeq \sum_{\beta\in\hat{C}_1}^{\oplus} \overline{\beta}\times \operatorname{ind}_{C_1\uparrow G_1} \beta .$$

Therefore by (A.10)

$$\sum_{\gamma\in\hat{G}_1}^{\oplus} \overline{\gamma}\big|_{C_1}\times\gamma \simeq \sum_{\beta\in\hat{C}_1}^{\oplus} \overline{\beta}\times \operatorname{ind}_{C_1\uparrow G_1} \beta$$

which proves Theorem A.3. Thus Theorem A.2 and Theorem A.3 are equivalent.

<u>Remark</u>: One knows that $\mathrm{ind}_{G_1^{\Delta}\uparrow G_1\times G_1} 1_{G_1^{\Delta}}$ is unitarily equivalent to the two-sided regular representation of $G_1 \times G_1$ on $L^2(G_1)$; see [27]. In view of this Proposition A.8 and Proposition 4.3.2 are entirely analogous. Similarly (A.10) is entirely analogous to Corollary 4.3.3.

REFERENCES

[1] N. Anh, Restriction of the Principal Series of SL(n, ℂ) to Some Reductive Subgroups, Pacific Journal of Mathematics, vol. 38, no. 2, 1971.

[2] R. Beals, Operators in Function Spaces which Commute with Multiplications, Duke Mathematical Journal, vol. 35, no. 2, 1968.

[3] R. Blattner, On Induced Representations, American Journal of Mathematics, vol. 83 (1961), 79-98.

[4] __________, On a Theorem of G. Mackey, Bulletin of the American Mathematical Society, vol. 68 (1962), 585-587.

[5] __________, Positive Definite Measures, Proceedings of the American Mathematical Society, vol. 14 (1963), 423-428.

[6] N. Bourbaki, Éléments de Mathématiques, Livre VI, Intégration, 1961, Paris.

[7] __________, (Seminaire), Analyse Spectrale et Théorème de Prédiction Statistique de Weiner, expose 218 (1961), Paris.

[8] F. Bruhat, Sur les Représentations Induites des Groups de Lie, Bulletin of the Society of Mathematics, France, vol. 84 (1956), 97-205.

[9] __________, Lectures on Lie Groups and Representations of Locally Compact Groups, Tata Institute of Fundamental Research, 1958, Bombay.

[10] C. Chevalley, Theory of Lie Groups, Princeton University Press, 1946, New Jersey.

[11] J. Dixmier, Les C*-algèbres et leurs Representations, Gauthier-Villars, Paris, 1964.

[12] I. Gelfand and M. Graev, Geometry of Homogeneous Spaces, Representations of Groups in Homogeneous Spaces and Related Questions of Integral Geometry, American Mathematical Society Translations, ser. 2, vol. 37.

[13] I. Gelfand and M. Naimark, Unitäre Darstellungen der Klassischen Gruppen, Akademie Verlag, Berlin, 1957.

[14] K. Gross, The Dual of a Parabolic Subgroup and a Degenerate Principal Series of $Sp(n, \mathbb{C})$, American Journal of Mathematics, vol. 93, no. 2, 1971.

[15] S. Helgason, Differential Geometry and Symmetric Spaces, Academic Press, New York, 1962.

[16] E. Hewitt and K. Ross, Abstract Harmonic Analysis, vol. 1, Springer-Verlag Berlin-Heidelberg-New York, 1963

[17] N. Jacobson, Lie Algebras, Interscience Publishers, New York, 1966.

[18] B. Kostant, On the Existence and Irreducibility of Certain Series of Representations, Bulletin of the American Mathematical Society, vol. 75 (1969), 627-642.

[19] R. Kunze and E. Stein, Uniformly Bounded Representations III, American Journal of Mathematics, vol. 89 (1967), 385-442.

[20] L. Loomis, Positive Definite Functions and Induced Representations, Duke Mathematical Journal, vol. 27 (1960), 569-580.

[21] P. Halmos, The Decomposition of Measures, Duke Mathematical Journal, vol. 8 (1941), 386-392.

[22] Harish-Chandra, Representations of Semi-Simple Lie Groups III, Transactions of the American Mathematical Society, vol. 76 (1954), 234-253.

[23] ______________, The Plancherel Formula for Complex Semi-Simple Lie Groups, Transactions of the American Mathematical Society, vol. 76 (1954), 485-528.

[24] ______________, On a Lemma of F. Bruhat, Journal de Mathématiques Pures et Appliquees, vol. 35 (1956), 203-210.

[25] G. Mackey, Induced Representations of Locally Compact Groups I, Annals of Mathematics (2)vol. 55 (1952), 101-139.

[26] ________, Induced Representations of Locally Compact Groups II, Annals of Mathematics (2)vol. 58 (1953), 193-221.

[27] ________, The Theory of Group Representations, Lecture notes, University of Chicago, 1955.

[28] ________, Imprimitivity for Representations of Locally Compact Groups I, Proceedings of the National Academy of Sciences U.S.A., vol. 35 (1949), 537-545.

[29] M. Naimark, Normed Rings, P. Noordhoff N. V.-Groningen, The Netherlands, 1964.

[30] ________, Decomposition of the Tensor Product of Irreducible Representations of a Proper Lorentz Group into Irreducible Representations, American Mathematical Society Translations, ser. 2, vol. 36.

[31] K. R. Parthasarathy, R. Ranga-Rao, and V. S. Varadarajan, Representations of Complex Semi-Simple Lie Groups and Lie Algebras, Annals of Mathematics (2)vol. 85 (1967), 383-429.

[32] J. Serre, Algèbres de Lie Semi-Simples Complexes, W. A. Benjamin, Incorporated, New York, 1966.

[33] N. Wallach, Cyclic Vectors and Irreducibility for Principal Series Representations, Transactions of the American Mathematical Society, vol. 158 (1971), 107-113.

[34] G. Warner, Harmonic Analysis on Semi-Simple Lie Groups, Springer-Verlag, Berlin-Heidelberg-New York, vols. 1 and 2.

[35] E. Wilson, Uniformly Bounded Representations of Semi-Simple Lie Groups, Thesis, Washington University, St. Louis, Missouri

[36] D. P. Zelobenko, Analysis of Irreducibility in the Class of Elementary Representations of a Complex Semi-Simple Lie Group, Math. USSR-Izv. 2 (1968), 105-128.

Vol. 215: P. Antonelli, D. Burghelea and P. J. Kahn, The Concordance-Homotopy Groups of Geometric Automorphism Groups. X, 140 pages. 1971. DM 16,-

Vol. 216: H. Maaß, Siegel's Modular Forms and Dirichlet Series. VII, 328 pages. 1971. DM 20,-

Vol. 217: T. J. Jech, Lectures in Set Theory with Particular Emphasis on the Method of Forcing. V, 137 pages. 1971. DM 16,-

Vol. 218: C. P. Schnorr, Zufälligkeit und Wahrscheinlichkeit. IV, 212 Seiten. 1971. DM 20,-

Vol. 219: N. L. Alling and N. Greenleaf, Foundations of the Theory of Klein Surfaces. IX, 117 pages. 1971. DM 16,-

Vol. 220: W. A. Coppel, Disconjugacy. V, 148 pages. 1971. DM 16,-

Vol. 221: P. Gabriel und F. Ulmer, Lokal präsentierbare Kategorien. V, 200 Seiten. 1971. DM 18,-

Vol. 222: C. Meghea, Compactification des Espaces Harmoniques. III, 108 pages. 1971. DM 16,-

Vol. 223: U. Felgner, Models of ZF-Set Theory. VI, 173 pages. 1971. DM 16,-

Vol. 224: Revètements Etales et Groupe Fondamental. (SGA 1). Dirigé par A. Grothendieck XXII, 447 pages. 1971. DM 30,-

Vol. 225: Théorie des Intersections et Théorème de Riemann-Roch. (SGA 6). Dirigé par P. Berthelot, A. Grothendieck et L. Illusie. XII, 700 pages. 1971. DM 40,-

Vol. 226: Seminar on Potential Theory, II. Edited by H. Bauer. IV, 170 pages. 1971. DM 18,-

Vol. 227: H. L. Montgomery, Topics in Multiplicative Number Theory. IX, 178 pages. 1971. DM 18,-

Vol. 228: Conference on Applications of Numerical Analysis. Edited by J. Ll. Morris. X, 358 pages. 1971. DM 26,-

Vol. 229: J. Väisälä, Lectures on n-Dimensional Quasiconformal Mappings. XIV, 144 pages. 1971. DM 16,-

Vol. 230: L. Waelbroeck, Topological Vector Spaces and Algebras. VII, 158 pages. 1971. DM 16,-

Vol. 231: H. Reiter, L^1-Algebras and Segal Algebras. XI, 113 pages. 1971. DM 16,-

Vol. 232: T. H. Ganelius, Tauberian Remainder Theorems. VI, 75 pages. 1971. DM 16,-

Vol. 233: C. P. Tsokos and W. J. Padgett. Random Integral Equations with Applications to stochastic Systems. VII, 174 pages. 1971. DM 18,-

Vol. 234: A. Andreotti and W. Stoll. Analytic and Algebraic Dependence of Meromorphic Functions. III, 390 pages. 1971. DM 26,-

Vol. 235: Global Differentiable Dynamics. Edited by O. Hájek, A. J. Lohwater, and R. McCann. X, 140 pages. 1971. DM 16,-

Vol. 236: M. Barr, P. A. Grillet, and D. H. van Osdol. Exact Categories and Categories of Sheaves. VII, 239 pages. 1971. DM 20,-

Vol. 237: B. Stenström, Rings and Modules of Quotients. VII, 136 pages. 1971. DM 16,-

Vol. 238: Der kanonische Modul eines Cohen-Macaulay-Rings. Herausgegeben von Jürgen Herzog und Ernst Kunz. VI, 103 Seiten. 1971. DM 16,-

Vol. 239: L. Illusie, Complexe Cotangent et Déformations I. XV, 355 pages. 1971. DM 26,-

Vol. 240: A. Kerber, Representations of Permutation Groups I. VII, 192 pages. 1971. DM 18,-

Vol. 241: S. Kaneyuki, Homogeneous Bounded Domains and Siegel Domains. V, 89 pages. 1971. DM 16,-

Vol. 242: R. R. Coifman et G. Weiss, Analyse Harmonique Non-Commutative sur Certains Espaces. V, 160 pages. 1971. DM 16,-

Vol. 243: Japan-United States Seminar on Ordinary Differential and Functional Equations. Edited by M. Urabe. VIII, 332 pages. 1971. DM 26,-

Vol. 244: Séminaire Bourbaki - vol. 1970/71. Exposés 382-399. IV, 356 pages. 1971. DM 26,-

Vol. 245: D. E. Cohen, Groups of Cohomological Dimension One. V, 99 pages. 1972. DM 16,-

Vol. 246: Lectures on Rings and Modules. Tulane University Ring and Operator Theory Year, 1970-1971. Volume I. X, 661 pages. 1972. DM 40,-

Vol. 247: Lectures on Operator Algebras. Tulane University Ring and Operator Theory Year, 1970-1971. Volume II. XI, 786 pages. 1972. DM 40,-

Vol. 248: Lectures on the Applications of Sheaves to Ring Theory. Tulane University Ring and Operator Theory Year, 1970-1971. Volume III. VIII, 315 pages. 1971. DM 26,-

Vol. 249: Symposium on Algebraic Topology. Edited by P. J. Hilton. VII, 111 pages. 1971. DM 16,-

Vol. 250: B. Jónsson, Topics in Universal Algebra. VI, 220 pages. 1972. DM 20,-

Vol. 251: The Theory of Arithmetic Functions. Edited by A. A. Gioia and D. L. Goldsmith VI, 287 pages. 1972. DM 24,-

Vol. 252: D. A. Stone, Stratified Polyhedra. IX, 193 pages. 1972. DM 18,-

Vol. 253: V. Komkov, Optimal Control Theory for the Damping of Vibrations of Simple Elastic Systems. V, 240 pages. 1972. DM 20,-

Vol. 254: C. U. Jensen, Les Foncteurs Dérivés de $\varprojlim$ et leurs Applications en Théorie des Modules. V, 103 pages. 1972. DM 16,-

Vol. 255: Conference in Mathematical Logic - London '70. Edited by W. Hodges. VIII, 351 pages. 1972. DM 26,-

Vol. 256: C. A. Berenstein and M. A. Dostal, Analytically Uniform Spaces and their Applications to Convolution Equations. VII, 130 pages. 1972. DM 16,-

Vol. 257: R. B. Holmes, A Course on Optimization and Best Approximation. VIII, 233 pages. 1972. DM 20,-

Vol. 258: Séminaire de Probabilités VI. Edited by P. A. Meyer. VI, 253 pages. 1972. DM 22,-

Vol. 259: N. Moulis, Structures de Fredholm sur les Variétés Hilbertiennes. V, 123 pages. 1972. DM 16,-

Vol. 260: R. Godement and H. Jacquet, Zeta Functions of Simple Algebras. IX, 188 pages. 1972. DM 18,-

Vol. 261: A. Guichardet, Symmetric Hilbert Spaces and Related Topics. V, 197 pages. 1972. DM 18,-

Vol. 262: H. G. Zimmer, Computational Problems, Methods, and Results in Algebraic Number Theory. V, 103 pages. 1972. DM 16,-

Vol. 263: T. Parthasarathy, Selection Theorems and their Applications. VII, 101 pages. 1972. DM 16,-

Vol. 264: W. Messing, The Crystals Associated to Barsotti-Tate Groups: With Applications to Abelian Schemes. III, 190 pages. 1972. DM 18,-

Vol. 265: N. Saavedra Rivano, Catégories Tannakiennes. II, 418 pages. 1972. DM 26,-

Vol. 266: Conference on Harmonic Analysis. Edited by D. Gulick and R. L. Lipsman. VI, 323 pages. 1972. DM 24,-

Vol. 267: Numerische Lösung nichtlinearer partieller Differential- und Integro-Differentialgleichungen. Herausgegeben von R. Ansorge und W. Törnig, VI, 339 Seiten. 1972. DM 26,-

Vol. 268: C. G. Simader, On Dirichlet's Boundary Value Problem. IV, 238 pages. 1972. DM 20,-

Vol. 269: Théorie des Topos et Cohomologie Etale des Schémas. (SGA 4). Dirigé par M. Artin, A. Grothendieck et J. L. Verdier. XIX, 525 pages. 1972. DM 50,-

Vol. 270: Théorie des Topos et Cohomologie Etale des Schémas. Tome 2. (SGA 4). Dirigé par M. Artin, A. Grothendieck et J. L. Verdier. V, 418 pages. 1972. DM 50,-

Vol. 271: J. P. May, The Geometry of Iterated Loop Spaces. IX, 175 pages. 1972. DM 18,-

Vol. 272: K. R. Parthasarathy and K. Schmidt, Positive Definite Kernels, Continuous Tensor Products, and Central Limit Theorems of Probability Theory. VI, 107 pages. 1972. DM 16,-

Vol. 273: U. Seip, Kompakt erzeugte Vektorräume und Analysis. IX, 119 Seiten. 1972. DM 16,-

Vol. 274: Toposes, Algebraic Geometry and Logic. Edited by. F. W. Lawvere. VI, 189 pages. 1972. DM 18,-

Vol. 275: Séminaire Pierre Lelong (Analyse) Année 1970-1971. VI, 181 pages. 1972. DM 18,-

Vol. 276: A. Borel, Représentations de Groupes Localement Compacts. V, 98 pages. 1972. DM 16,-

Vol. 277: Séminaire Banach. Edité par C. Houzel. VII, 229 pages. 1972. DM 20,-

Vol. 278: H. Jacquet, Automorphic Forms on GL(2). Part II. XIII, 142 pages. 1972. DM 16,–

Vol. 279: R. Bott, S. Gitler and I. M. James, Lectures on Algebraic and Differential Topology. V, 174 pages. 1972. DM 18,–

Vol. 280: Conference on the Theory of Ordinary and Partial Differential Equations. Edited by W. N. Everitt and B. D. Sleeman. XV, 367 pages. 1972. DM 26,–

Vol. 281: Coherence in Categories. Edited by S. Mac Lane. VII, 235 pages. 1972. DM 20,–

Vol. 282: W. Klingenberg und P. Flaschel, Riemannsche Hilbertmannigfaltigkeiten. Periodische Geodätische. VII, 211 Seiten. 1972. DM 20,–

Vol. 283: L. Illusie, Complexe Cotangent et Déformations II. VII, 304 pages. 1972. DM 24,–

Vol. 284: P. A. Meyer, Martingales and Stochastic Integrals I. VI, 89 pages. 1972. DM 16,–

Vol. 285: P. de la Harpe, Classical Banach-Lie Algebras and Banach-Lie Groups of Operators in Hilbert Space. III, 160 pages. 1972. DM 16,–

Vol. 286: S. Murakami, On Automorphisms of Siegel Domains. V, 95 pages. 1972. DM 16,–

Vol. 287: Hyperfunctions and Pseudo-Differential Equations. Edited by H. Komatsu. VII, 529 pages. 1973. DM 36,–

Vol. 288: Groupes de Monodromie en Géométrie Algébrique. (SGA 7 I). Dirigé par A. Grothendieck. IX, 523 pages. 1972. DM 50,–

Vol. 289: B. Fuglede, Finely Harmonic Functions. III, 188. 1972. DM 18,–

Vol. 290: D. B. Zagier, Equivariant Pontrjagin Classes and Applications to Orbit Spaces. IX, 130 pages. 1972. DM 16,–

Vol. 291: P. Orlik, Seifert Manifolds. VIII, 155 pages. 1972. DM 16,–

Vol. 292: W. D. Wallis, A. P. Street and J. S. Wallis, Combinatorics: Room Squares, Sum-Free Sets, Hadamard Matrices. V, 508 pages. 1972. DM 50,–

Vol. 293: R. A. DeVore, The Approximation of Continuous Functions by Positive Linear Operators. VIII, 289 pages. 1972. DM 24,–

Vol. 294: Stability of Stochastic Dynamical Systems. Edited by R. F. Curtain. IX, 332 pages. 1972. DM 26,–

Vol. 295: C. Dellacherie, Ensembles Analytiques, Capacités, Mesures de Hausdorff. XII, 123 pages. 1972. DM 16,–

Vol. 296: Probability and Information Theory II. Edited by M. Behara, K. Krickeberg and J. Wolfowitz. V, 223 pages. 1973. DM 20,–

Vol. 297: J. Garnett, Analytic Capacity and Measure. IV, 138 pages. 1972. DM 16,–

Vol. 298: Proceedings of the Second Conference on Compact Transformation Groups. Part 1. XIII, 453 pages. 1972. DM 32,–

Vol. 299: Proceedings of the Second Conference on Compact Transformation Groups. Part 2. XIV, 327 pages. 1972. DM 26,–

Vol. 300: P. Eymard, Moyennes Invariantes et Représentations Unitaires. II, 113 pages. 1972. DM 16,–

Vol. 301: F. Pittnauer, Vorlesungen über asymptotische Reihen. VI, 186 Seiten. 1972. DM 18,–

Vol. 302: M. Demazure, Lectures on p-Divisible Groups. V, 98 pages. 1972. DM 16,–

Vol. 303: Graph Theory and Applications. Edited by Y. Alavi, D. R. Lick and A. T. White. IX, 329 pages. 1972. DM 26,–

Vol. 304: A. K. Bousfield and D. M. Kan, Homotopy Limits, Completions and Localizations. V, 348 pages. 1972. DM 26,–

Vol. 305: Théorie des Topos et Cohomologie Etale des Schémas. Tome 3. (SGA 4). Dirigé par M. Artin, A. Grothendieck et J. L. Verdier. VI, 640 pages. 1973. DM 50,–

Vol. 306: H. Luckhardt, Extensional Gödel Functional Interpretation. VI, 161 pages. 1973. DM 18,–

Vol. 307: J. L. Bretagnolle, S. D. Chatterji et P.-A. Meyer, Ecole d'été de Probabilités: Processus Stochastiques. VI, 198 pages. 1973. DM 20,–

Vol. 308: D. Knutson, λ-Rings and the Representation Theory of the Symmetric Group. IV, 203 pages. 1973. DM 20,–

Vol. 309: D. H. Sattinger, Topics in Stability and Bifurcation Theory. VI, 190 pages. 1973. DM 18,–

Vol. 310: B. Iversen, Generic Local Structure of the Morphisms in Commutative Algebra. IV, 108 pages. 1973. DM 16,–

Vol. 311: Conference on Commutative Algebra. Edited by J. W. Brewer and E. A. Rutter. VII, 251 pages. 1973. DM 22,–

Vol. 312: Symposium on Ordinary Differential Equations. Edited by W. A. Harris, Jr. and Y. Sibuya. VIII, 204 pages. 1973. DM 22,–

Vol. 313: K. Jörgens and J. Weidmann, Spectral Properties of Hamiltonian Operators. III, 140 pages. 1973. DM 16,–

Vol. 314: M. Deuring, Lectures on the Theory of Algebraic Functions of One Variable. VI, 151 pages. 1973. DM 16,–

Vol. 315: K. Bichteler, Integration Theory (with Special Attention to Vector Measures). VI, 357 pages. 1973. DM 26,–

Vol. 316: Symposium on Non-Well-Posed Problems and Logarithmic Convexity. Edited by R. J. Knops. V, 176 pages. 1973. DM 18,–

Vol. 317: Séminaire Bourbaki – vol. 1971/72. Exposés 400–417. IV, 361 pages. 1973. DM 26,–

Vol. 318: Recent Advances in Topological Dynamics. Edited by A. Beck, VIII, 285 pages. 1973. DM 24,–

Vol. 319: Conference on Group Theory. Edited by R. W. Gatterdam and K. W. Weston. V, 188 pages. 1973. DM 18,–

Vol. 320: Modular Functions of One Variable I. Edited by W. Kuyk. V, 195 pages. 1973. DM 18,–

Vol. 321: Séminaire de Probabilités VII. Edité par P. A. Meyer. VI, 322 pages. 1973. DM 26,–

Vol. 322: Nonlinear Problems in the Physical Sciences and Biology. Edited by I. Stakgold, D. D. Joseph and D. H. Sattinger. VIII, 357 pages. 1973. DM 26,–

Vol. 323: J. L. Lions, Perturbations Singulières dans les Problèmes aux Limites et en Contrôle Optimal. XII, 645 pages. 1973. DM 42,–

Vol. 324: K. Kreith, Oscillation Theory. VI, 109 pages. 1973. DM 16,–

Vol. 325: Ch.-Ch. Chou, La Transformation de Fourier Complexe et L'Equation de Convolution. IX, 137 pages. 1973. DM 16,–

Vol. 326: A. Robert, Elliptic Curves. VIII, 264 pages. 1973. DM 22,–

Vol. 327: E. Matlis, 1-Dimensional Cohen-Macaulay Rings. XII, 157 pages. 1973. DM 18,–

Vol. 328: J. R. Büchi and D. Siefkes, The Monadic Second Order Theory of All Countable Ordinals. VI, 217 pages. 1973. DM 20,–

Vol. 329: W. Trebels, Multipliers for (C, α)-Bounded Fourier Expansions in Banach Spaces and Approximation Theory. VII, 103 pages. 1973. DM 16,–

Vol. 330: Proceedings of the Second Japan-USSR Symposium on Probability Theory. Edited by G. Maruyama and Yu. V. Prokhorov. VI, 550 pages. 1973. DM 36,–

Vol. 331: Summer School on Topological Vector Spaces. Edited by L. Waelbroeck. VI, 226 pages. 1973. DM 20,–

Vol. 332: Séminaire Pierre Lelong (Analyse) Année 1971-1972. V, 131 pages. 1973. DM 16,–

Vol. 333: Numerische, insbesondere approximationstheoretische Behandlung von Funktionalgleichungen. Herausgegeben von R. Ansorge und W. Törnig. VI, 296 Seiten. 1973. DM 24,–

Vol. 334: F. Schweiger, The Metrical Theory of Jacobi-Perron Algorithm. V, 111 pages. 1973. DM 16,–

Vol. 335: H. Huck, R. Roitzsch, U. Simon, W. Vortisch, R. Walden, B. Wegner und W. Wendland, Beweismethoden der Differentialgeometrie im Großen. IX, 159 Seiten. 1973. DM 18,–

Vol. 336: L'Analyse Harmonique dans le Domaine Complexe. Edité par E. J. Akutowicz. VIII, 169 pages. 1973. DM 18,–

Vol. 337: Cambridge Summer School in Mathematical Logic. Edited by A. R. D. Mathias and H. Rogers. IX, 660 pages. 1973. DM 42,–

Vol: 338: J. Lindenstrauss and L. Tzafriri, Classical Banach Spaces. IX, 243 pages. 1973. DM 22,–

Vol. 339: G. Kempf, F. Knudsen, D. Mumford and B. Saint-Donat, Toroidal Embeddings I. VIII, 209 pages. 1973. DM 20,–

Vol. 340: Groupes de Monodromie en Géométrie Algébrique. (SGA 7 II). Par P. Deligne et N. Katz. X, 438 pages. 1973. DM 40,–

Vol. 341: Algebraic K-Theory I, Higher K-Theories. Edited by H. Bass. XV, 335 pages. 1973. DM 26,–

Vol. 342: Algebraic K-Theory II, "Classical" Algebraic K-Theory, and Connections with Arithmetic. Edited by H. Bass. XV, 527 pages. 1973. DM 36,–